benefits of a more varied workforce. It goes beyond simply identifying issues by providing real-world strategies that industry leaders, educational institutions, and policymakers can implement. The book takes a two-fold approach: it first exposes the challenges that have kept marginalized communities out of transportation careers, from stereotypes to educational barriers. Then, it sets forth a detailed plan for overcoming these hurdles through mentorship programs, scholarship initiatives, and workplace reforms. Rather than just diagnosing problems, *Bridging the Gap* lays out a step-by-step roadmap for meaningful change. Its comprehensive framework positions it as a valuable resource for anyone committed to enhancing both the effectiveness and social responsibility of the transportation industry. Whether you're an individual looking to contribute or an organization aiming to update its practices, this book serves as an excellent guide for driving positive change.

—MICHAEL L. WESTRAY, JR.
CEO/Lead Consultant, Optimus Business Solutions, LLC

I have had the privilege of reviewing Esther Franklin's groundbreaking work, *Bridging the Gap: Breaking Barriers and Building Pathways to Transportation Careers*. This book is not just a testament to the pressing need for diversity in the transportation industry but also a beacon of hope for countless individuals seeking meaningful careers. Through meticulous research and compelling narratives, Esther has illuminated the profound impact that individuals from diverse backgrounds can have on the transportation sector. Her emphasis on inclusivity, representation, and the transformative power of these careers is both timely and essential. The stories of triumph, resilience, and determination showcased in *Bridging the Gap* serve as a rallying call for a more equitable and representative transportation workforce. It is a clarion call for industry leaders, policymakers, and educational institutions to come together and pave the way for a brighter, more inclusive future. I wholeheartedly endorse *Bridging the Gap* and commend Esther Franklin for her dedication to driving change, breaking barriers, and championing inclusivity in the transportation industry. This book is a must read for anyone passionate about fostering diversity, innovation, and progress.

—KYRON ROBINSON
ProRank Managing Partner, DBE SSC Program Manager,
ProRank Business Solutions

Esther Franklin has a marvelous skill set associated with the country's important infrastructure. She has taken on complicated projects that other people shy away from. My career in title insurance began in 1989, and I learn new things from Esther all the time due to the interesting and unique assignments that she embraces. It is great that she has written this book to bring to light all that this field has to offer.

—JUDY NEMETH
Agency Manager, Old Republic Title |
Old Republic Insurance Group

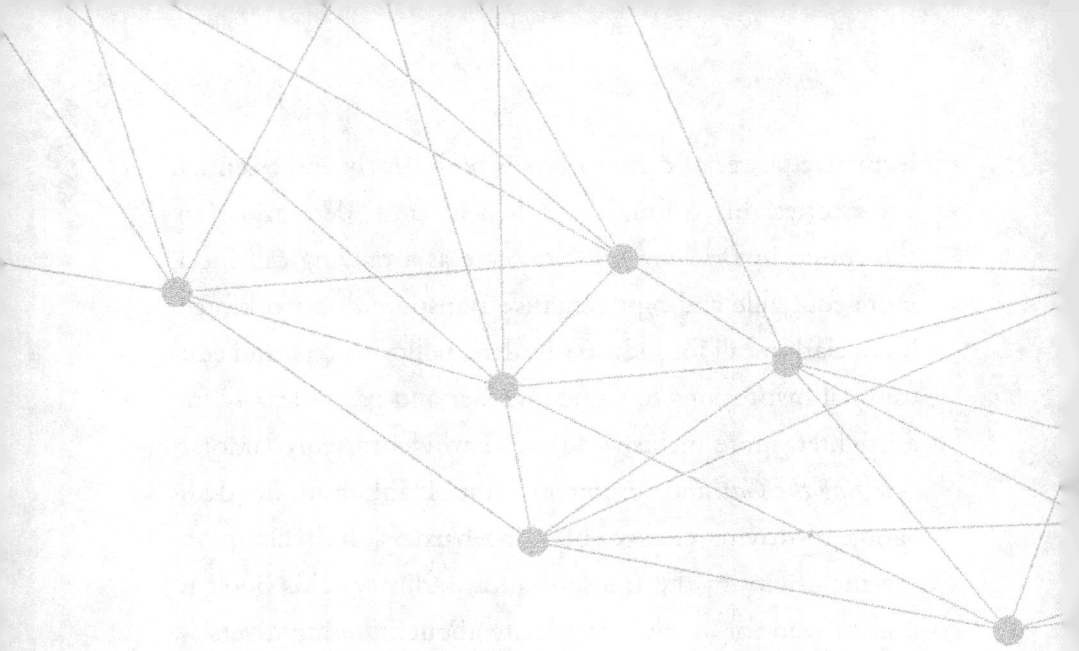

BRIDGING THE GAP

BRIDGING THE GAP

BREAKING BARRIERS AND BUILDING PATHWAYS
TO TRANSPORTATION CAREERS

ESTHER FRANKLIN

Forbes | Books

Published by Forbes Books, Charleston, South Carolina.
An imprint of Advantage Media Group.

Forbes Books is a registered trademark, and the Forbes Books colophon is a trademark of Forbes Media, LLC.

Printed in the United States of America.

10 9 8 7 6 5 4 3 2 1

ISBN: 979-8-88750-456-8 (Hardcover)
ISBN: 979-8-88750-457-5 (eBook)

Library of Congress Control Number: 2024926681

Cover and Layout design by Matthew Morse.

This custom publication is intended to provide accurate information and the opinions of the author in regard to the subject matter covered. It is sold with the understanding that the publisher, Forbes Books, is not engaged in rendering legal, financial, or professional services of any kind. If legal advice or other expert assistance is required, the reader is advised to seek the services of a competent professional.

Since 1917, Forbes has remained steadfast in its mission to serve as the defining voice of entrepreneurial capitalism. Forbes Books, launched in 2016 through a partnership with Advantage Media, furthers that aim by helping business and thought leaders bring their stories, passion, and knowledge to the forefront in custom books. Opinions expressed by Forbes Books authors are their own. To be considered for publication, please visit books.Forbes.com.

I would like to dedicate this book to my three children,
Martae, Gerard, and Geordan, for being the best part of my life.
Your births were high points in my life.
Giving you life gave me more life.

Martae, you were born soon after I graduated from Duquesne High School. The class of 1996 was Duquesne High School's one hundredth commencement; this was the start of my journey.

Gerard, you were born right after my college graduation from the University of Pittsburgh. The class of 2000 was the largest graduating class in University of Pittsburgh history; this was my discovery phase.

Geordan, my one and only daughter, you were born soon after the death of your father; this was the turning point in my life where I decided to be the best version of me for you all.

Each one of you has purpose, and I am grateful to be your mother. While I pursued the education I needed to build businesses, support my family, and create a legacy for my children's future, Milton Hershey School delivered a quality education to all my children. I'm truly grateful my family was able to benefit from the school's mission and commitment to provide a nurturing environment for their students. Many thanks to the teachers and administrative staff who take that oath seriously.

CONTENTS

PART 1:
A BUILDING BOOM IS COMING

PART 2:
CHOOSING YOUR LANE:
THE NON-REAL ESTATE SECTOR

PART 3:
CHOOSING YOUR LANE:
THE REAL ESTATE SECTOR

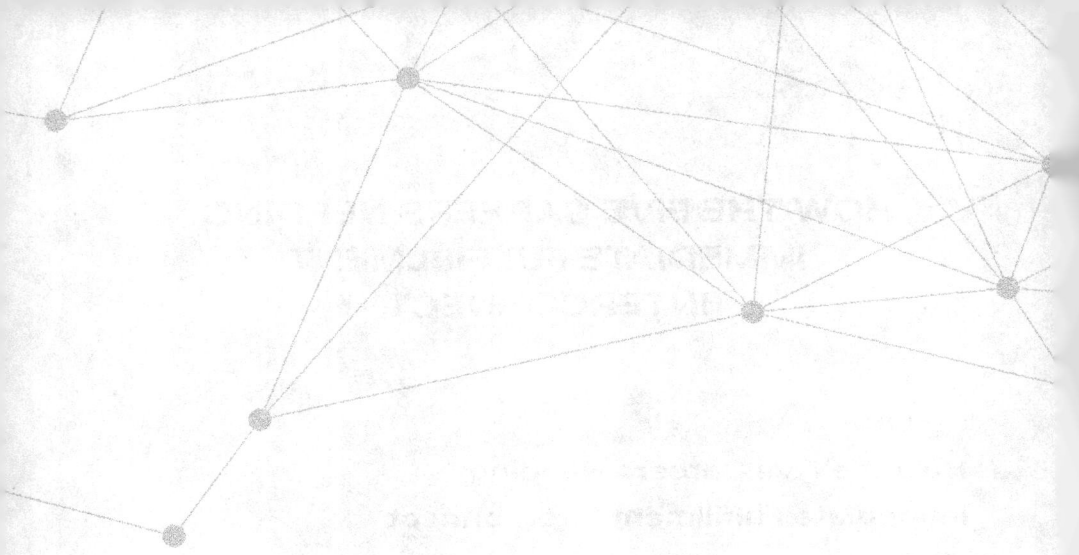

ACKNOWLEDGMENTS

Without the contributors to our careers in need, this book truly would not have been possible. Each professional showed an amazing perceptivity in their field, and my team and I enjoyed broadening our knowledge of their specific project perspectives.

Engineer Jared Green has an amazing way of answering a question by asking more questions that somehow produce the answer.

Architect Pascale Sablan's brilliance is uniquely complemented by grace, wisdom, and care for her communities.

Right-of-way professional Chuck Latus is committed to a straightforward, fair approach to a difficult job, and it is to be greatly admired.

Appraiser Derek Molen exhibits wisdom beyond his years; his insight and patience in bringing my team up to speed was greatly appreciated.

It became especially intriguing when the following four gaps in our collective industries were recognized, noted, and expressed without prodding from my team:

- Each expected the highway infrastructure bill to produce jobs, saw it as very much needed, and felt their fields would need next-generation candidates.

- Each understood equality issues and how the Bipartisan Infrastructure Law endeavors to relieve unfair past practices and move forward in a more equitable manner.

- Each expressed a hesitancy about allowing AI to have too much input into their job fields.

- Each understood the reliance on teamwork within their respective fields in order to bring a project to completion.

A special thank-you to each of you for sharing your knowledge, experiences, and unique perspectives. Your collective wisdom has brought depth and authenticity to the subject matter. Your attention to detail, precision, and commitment to excellence have made this book not just a product of my own efforts but a collaborative masterpiece we can all be proud of. As we move forward, I am certain that the impact of our collective work will be felt far and wide.

INTRODUCTION

Most readers likely already know that our economy is experiencing a job-versus-worker crisis. Fast-food jobs are available on every corner, but jobs that can support a robust lifestyle, careers that give meaning to the day-to-day, are severely lacking. Americans want good-paying jobs with benefits and stability, yet they aren't as easily found when the actual job hunt begins. Many people feel the situation is hopeless.

There is an industry, however, where these criteria can be met. Quite possibly, the jobs available in this field can exceed your expectations. It's the transportation industry. This industry consists of systems of roadways, airways, railways, waterways, bridges—even outer space—that work together to facilitate the movement of goods and people. Some parts of the overall field are in desperate need for determined, take-charge individuals to learn proper procedure, follow through on it, and perform their tasks accordingly. The best part about the current and future opportunities expected is that the industry is so broad, opportunities to find your career match abound. Higher education is not necessarily required in some cases, and the pay can be substantial.

THE TRANSPORTATION INDUSTRY

So what exactly is the transportation industry? What does it do? What types of jobs are available?

Just as a light suddenly switching on will illuminate every corner of a dark room, our transportation system reaches the smallest towns and the biggest cities with goods and services. To achieve this, the giant working system of the US transportation infrastructure is managed by millions of jobs. Every product, service, or change in the overall economy starts a chain reaction of needed workers to make that product or service a reality.

For instance, an energy shortage is currently resulting in high gas prices. This brings electric vehicles to the forefront. Many consumers who previously were uninterested in EVs are now turning to them as an alternative to gasoline-powered engines. Just this one trend toward EVs provided jobs for 7.8 million people in 2021.

"US electric vehicle jobs led 2021 energy sector growth: Overall, the sector rose by 4 percent from 2020 to 2021, outpacing overall US employment, which grew 2.8 percent in the same period. The energy sector added more than 300,000 jobs, increasing the total number of energy jobs from 7.5 million in 2020 to more than 7.8 million in 2021."[1]

As we approach 2030, the Economic Policy Institute estimates "the shift to all-electric vehicles could create over 150,000 jobs by 2030."[2]

---

1   Michelle Lewis, "US Electric Vehicle Jobs Exploded in 2021, Clean Energy Jobs Grew—and Fossil Fuel Jobs Shrunk, Electrek, June 28, 2022, https://electrek.co/2022/06/28/us-electric-vehicle-jobs-2021/#:~:text=Overall%20percent2C%20percent20the%20percent20sector%20percent20rose%20percent20by,than%20percent207.8%20percent20million%20percent20in%20percent202021.

2   Economic Policy Institute, News from EPI, "The Shift to All-Electric Vehicles Could Create over 150,000 Jobs by 2030—If Policymakers Make Smart Investments to Secure US Leadership in the Auto Sector, accessed September 12, 2023, https://www.epi.org/press/the-shift-to-all-electric-vehicles-could-create-over-150000-jobs-by-2030-if-policymakers-make-smart-investments-to-secure-u-s-leadership-in-the-auto-sector/.

Additionally, it's not just metal and bolt suppliers who prosper when there's a demand for new cars. There are glass, plastics, carpet, interior-fabrics, and seating suppliers—just to name a few—whose businesses and employees will thrive right along with the EV manufacturer.

Of course, newer cars mean newer amenities are expected, along with the electric engine. The modern automobile offers advanced conveniences, from trip planning to onboard navigation to onboard autopilot. Each one of these features will set further hundreds, if not thousands, of employment outcomes into motion. The advanced computer systems mean the various computer sectors will be called upon for programmers, data analysts, and cybersecurity specialists; safety experts; vehicle-device programmers and instructors at the customer level; and sales, customer service, and reception at the customer level. The list goes on and on, right down to the copywriter who edits the welcome letters for all those new home pages. Professionals from every education level and socio-economic circumstance are needed to fill these needs.

For those who have the ability to plan for a specific transportation career and gain education along those lines, there are jobs waiting for fulfillment right now, and the need will only increase with time. If schooling is not in your plan, it's not necessary to embark on a long path of education because whatever skills you currently have will almost certainly fit somewhere in the job boom being predicted. The whole of the transportation infrastructure is so multilayered that many fields, seemingly having nothing to do with transportation, funnel into its category at the end of the day.

Let's take a moderately sized audiovisual company, for instance. At first glance, AV technology has nothing to do with transportation. However, nothing happens in our society without photo and video documentation. So even if the DOT is not specifically posting audiovisual positions, AV experts can creatively market to this field of business. Just as manufacturing an electric vehicle creates a spiderweb effect of related work, the same will be true of multiple job arenas as they relate to the transportation industry.

INDUSTRY VERSUS INFRASTRUCTURE

When explaining how different jobs fit in the overall transportation industry, this question inevitably arises: What's the difference between the "transportation industry" and the "transportation infrastructure"? *National Geographic* puts out an ideal explanation:

"Transportation is the movement of goods and people from one place to another. In ancient times, people crafted simple boats out of logs, walked, rode animals and, later, devised wheeled vehicles to move from place to place. They used existing waterways or simple roads for transportation. Over time, people built more complex means of transportation. They learned how to harness various sources of power, such as wind, steam, and combustion, to move barges, ships, trains, automobiles, and airplanes. These new means of transportation required people to change their environments by building transportation infrastructure. Transportation infrastructure is the underlying system of public works designed to facilitate movement."[3]

So the transportation *industry* consists of companies or government entities that move people or goods by roadway, air, water, rail, or outer space. The transportation *infrastructure* is a system of public works that supports that industry with laws and standards of operation. Because we'll need to be well-versed in this information as

---

3   National Geographic Education, "Transportation Infrastructure," accessed September 12, 2023, https://education.nationalgeographic.org/resource/transportation-infrastructure.

we go, take a moment to become familiar with the official definitions of these infrastructure components:

Infrastructure Classifications

ROADS	"The US DOT's Federal Highway Administration (FHWA) classifies our Nation's urban and rural roadways by road function. Each function class is based on the type of service the road provides to the motoring public, and the designation is used for data and planning purposes. Design standards are tied to function class. Each class has a range of allowable lane widths, shoulder widths, curve radii, etc."[4]
INTERSTATE SYSTEM	"The Interstate System is the highest classification of roadways in the United States. These arterial roads provide the highest level of mobility and the highest speeds over the longest uninterrupted distance. Interstates nationwide usually have posted speeds between 55 and 75 mi/h."[5]
OTHER ARTERIALS	"Other Arterials include freeways, multilane highways, and other important roadways that supplement the Interstate System. They connect, as directly as practicable, the Nation's principal urbanized areas, cities, and industrial centers. Land access is limited. Posted speed limits on arterials usually range between 50 and 70 mi/h."[6]
COLLECTORS	"Collectors are major and minor roads that connect local roads and streets with arterials. Collectors provide less mobility than arterials at lower speeds and for shorter distances. They balance mobility with land access. The posted speed limit on collectors is usually between 35 and 55 mph."[7]
LOCAL ROADS	"Local roads provide limited mobility and are the primary access to residential areas, businesses, farms, and other local areas. Local roads, with posted speed limits usually between 20 and 45 mi/h, are the majority of roads in the US."[8]

---

4   US Department of Transportation, "Road Classifications," accessed September 12, 2023, https://safety.fhwa.dot.gov/speedmgt/data_facts/docs/rd_func_class_1_42.pdf.

5   US Dept. of Transportation, "Road Classifications."

6   US Dept. of Transportation, "Road Classifications."

7   US Dept. of Transportation, "Road Classifications."

8   US Dept. of Transportation, "Road Classifications."

Infrastructure Classifications

WATERWAYS	"A narrow area of water, such as a river or canal, that ships or boats can sail along."[9]
AIRWAYS	"A designated route along which airplanes fly from airport to airport; especially such a route equipped with navigational aids."[10]
RAILWAYS	"A permanent track composed of a line of parallel metal rails fixed to sleepers, for transport of passengers and goods in trains."[11]
BRIDGES	"Bridges and structures are key components of the nation's roadway network that provide transportation connectivity to safely cross features such as waterways, railways, roadways, and other obstacles ... The Office of Bridges and Structures provides a technical function to support the safety, stewardship, and oversight of over 610,000 highway bridges, more than 500 tunnels, and numerous other structures across the entire USA. Under the Federal-Aid Highway Program, FHWA annually distributes funding of approximately $7 billion to assist transportation agencies plan, design, build, repair, rehabilitate, and inspect such bridges and structures."[12]
OUTER SPACE	"Outer space is the region beyond a planet's atmosphere. For Earth, it begins about 100 kilometers (62 miles) above sea level. The line separating the atmosphere and outer space is called the Kármán line. The term outer space and the universe are roughly equivalent, except that outer space refers only to the area between planets, while the universe encompasses planets as well."[13]

9   *Cambridge Dictionary*, s.v. "waterway (*n*)" accessed September 12, 2023, https://dictionary.cambridge.org/us/dictionary/english/waterway.

10 *Merriam Webster Dictionary*, s.v. "airway (*n*)," accessed September 12, 2023, https://www.merriam-webster.com/dictionary/airway.

11 *Collins Dictionary*, s.v. "railway (*n*)," accessed September 12, 2023, https://www.collinsdictionary.com/us/dictionary/english/railway.

12 US Department of Transportation Federal Highway Administration, "Bridges and Structures," accessed September 12, 2023, https://www.fhwa.dot.gov/bridge/#:~:text=Bridges%20percent20and%20percent20structures%20percent20are%20percent20key,%20percent2C%20percent20%20roadways%20percent2C%20percent20and%20percent20other%20percent20obstacles.

13 Gavin Wright, "What Is Outer Space," WhatIs.com, accessed September 12, 2023, https://www.techtarget.com/whatis/definition/space#:~:text=Outer percent20space percent20is percent20the percent20region,is percent20called percent20the percent20K percentC3 percentA1rm percentC3 percentA1n percent20line.

It's truly difficult to convey how the many facets of movement work together. Each individual item created or service rendered brings on another trek through the manufacturing and delivery process. In each individual step, multiple jobs are created—just as we see happening today with EVs.

If all of this is true, however, why haven't we heard about and seen these opportunities before now? Quite frankly, because funding for transportation infrastructure upkeep and repair has been neglected by our national leadership.

CRISIS ON AMERICA'S ROADWAYS

Most motorists are painfully aware that inattention has put our US roads and highways into a serious state of disrepair. In some cases, they are literally crumbling—causing flat tires, motor vehicle accidents, and traffic jams.

During the boredom of slow-moving traffic, we see driver attention turning to phones for texting, conversations, even surfing the web. More than three thousand people die in distracted-driving accidents each year; distracted driving causes about four hundred thousand injuries annually and almost one million accidents per year.[14] Distracted driving behaviors intensify when road conditions are not safe to begin with.

A worse phenomenon of our time is road rage—or simply impatient driving. This is a very real and dangerous reaction to the stress of traffic. This driver is more prone to take dangerous chances, speed up at a yellow light rather than slow down, or weave in and

14   Pete Ortiz, "10 Distracted Driving Facts and Statistics—2023 UPDATE," House Grail, accessed September 12, 2023, https://housegrail.com/distracted-driving-facts-statistics/.

out of slower-moving traffic. This, of course, increases the chances of a serious collision.

Then, of course, there are the death statistics, the most crushing and often most unnecessary of all. Forbes Advisor reports, "Car accidents are annually responsible for approximately 1.3 million deaths worldwide, according to the World Health Organization (WHO). In the United States, the National Highway Traffic Safety Administration (NHTSA) projects there were an estimated 42,915 traffic fatalities in 2021, a 10.5 percent increase compared to 2020 and the highest annual percentage increase in the Fatality Analysis Reporting System's history."[15]

Much of this heartache could be resolved with proper attention to our infrastructure. According to the American Society of Civil Engineers, "Growing wear and tear on our nation's roads has left 43 percent of our public roadways in poor or mediocre condition, a number that has remained stagnant over the past several years; there is a water main break every two minutes, and an estimated 6 billion gallons of treated water lost each day in the US, enough to fill over 9,000 swimming pools; there are 30,000 miles of inventoried levees across the US, and an additional 10,000 miles of levees whose location and condition are unknown."[16]

Additionally, the American Road and Transportation Builders Association (ARTBA) made these alarming observations in their *2023 Bridge Report*: "There are 163.2 million daily crossings on 42,951 struc-

---

15   Shelby Simon, "How Many People Die from Car Accidents Each Year?," Forbes Advisor, updated October 10, 2022, https://www.forbes.com/advisor/legal/auto-accident/car-accident-deaths/#:~:text= More percent-20than percent2046 percent2C000 percent20people percent20die,12.4 percent20deaths percent20per percent20100 percent2C000 percent20inhabitants.

16   "Report Card for American Infrastructure, America's Infrastructure Scores a C-," accessed September 12, 2023, https://infrastructurereportcard.org/#:~:text=have percent20left percent2043 percent25 percent20of percent20our,over percent20the percent20past percent20several percent20years.

turally deficient US bridges in poor condition" and "The length of the 223,000 bridges in need of repair would stretch over 6,100 miles."[17]

This information is astounding, and the safety issue is what highlights the need for immediate attention. When a government fails to maintain its roadways, the citizenry are the ones who pay the price. So what can we do about it?

In this book, I'll show you that funding for highway construction and repairs has brought about unprecedented job opportunities. However, given the number of projects that will be undertaken across the country, there simply isn't adequate staffing in the different transportation-infrastructure-related disciplines to handle the influx of work.

How about getting down to the real reasons such a gap between work and available workers exists? We'll address some startling inequality issues head-on, with a mind toward repairing damages and avoiding the same mistakes in the future.

Let's talk plainly about what we're all interested in: money and jobs with great benefits packages. I'll show you jobs for all educational backgrounds, in clear-list format, so you can begin exploring careers that really suit you.

Are you a good fit for one of the five career positions we will highlight as "needing immediate fulfillment": engineering, architecture, right-of-way acquisition, appraisals, and title specialty? We'll gain a clear idea of those career paths with featured chapters from experts in their fields to help us get a realistic feel for the jobs.

First, however, let's try to get a picture of just how big the transportation industry is. Let's learn a bit about it, why it's been in such a crisis, and how the infrastructure of this industry connects the entire world.

---

17   American Road & Transportation Builders Association, *Bridge Report*, accessed September 12, 2023, https://artbabridgereport.org.

PART 1

A BUILDING BOOM IS COMING

1

CONNECTING THE WORLD

Mid-August of this year, just as commuters have gotten used to traveling without the interference of school buses, traffic will undoubtedly be on the rise again. Millions of back-to-school shoppers will enter the roadways in search of the most-current clothing styles and supplies. Everything from jeans to bookbags to pens to calculators will be stocked and waiting for these consumers.

It's a little-realized fact that none of these products would be on the shelves without an extensive labyrinth of seaways, airways, railways, roadways, and bridges. Without them, we couldn't move people or products any substantial distance. Food, clothing, and shelter would rely on the natural, individual skills each person brought to their household.

For instance, let's take just one pair of boys' blue jeans, purchased during that school-shopping traffic jam. Watch how the process of creating and delivering them involves every one of the movement platforms of the transportation infrastructure:

In-Store Shopping Process

The first step in processing those jeans will take place in Texas. Cotton fiber is grown, machine picked, and then separated from seeds and debris. Right off the bat, multiple companies are involved, and hundreds of employees have reliable work.

This product then travels across land via semitrucks and trains, employing hundreds in a completely different sector of transportation.

Ocean containers then ship the unfinished product to a textile mill in Bangladesh, creating a cascade of work not only for the residents of that nation but also for the shipping companies and customs personnel who must become involved.

In Bangladesh, numerous skilled workers spin the product into yarn.

Still more workers weave the yarn into denim fabric.

The fabric is transported back to the US via international waterways, creating another round of work for shipping and customs agencies.

More trains or trucks, requiring more workers, will transport the denim to a blue jeans producer in California.

California garment workers then dye, cut, and sew that denim into a pair of jeans.

In-Store Shopping Process

To produce those jeans, garment workers use thread, zippers, and buttons that also traveled internationally to their work floor, creating this pattern all over again with those products.

The finished jeans are packed up by a shipping team.

An eighteen-wheel semitruck driver transports them over America's highways, eventually dropping them off at a warehouse in middle America.

There, another untold number of workers will sort and store them.

Eventually yet another truck will move the jeans over another set of highways to retail stores all over the nation.

The consumer will travel by car, bus, subway, or bicycle to make commerce go around by purchasing that pair of jeans at his local mall.

Online shoppers cause that commerce to go around as well—make no mistake. All the same processing of the garment takes place, along with all the worldwide movement of product, but these consumers will order their new pairs of jeans online, and the remainder of the process will look like this:

Online Shipping Process

A courier service—FedEx, UPS, USPS, or Amazon Prime—will pick them up from a central warehouse.

The jeans will then be trucked through your neighborhood streets before being dropped off at your front door.

In the future, those jeans may arrive directly at your home via drone or self-driving car—paid for in cryptocurrency, of course.

It's apparent that the number of jobs this "simple" shipment of blue jeans creates is too numerous for us to count. Unfortunately, the transportation industry is lacking enough workers to competently take on several necessary positions that keep this process flowing. Why is that? There are two things happening simultaneously that are about to cause a job explosion: mass retirements and the $1 trillion Bipartisan Infrastructure Law (BIL). Let's take a look at each.

RETIREMENTS

Between the years of 1946 and 1967, a strong economy and family-centric values brought about the baby boom. Many of the Americans born during that nearly twenty-year period grew up to take positions in federal offices. They are now between the approximate ages of fifty-six and seventy-seven—the retirement years. This is creating a mass-exit problem in nearly every federal agency across the nation. The US Department of Transportation (DOT) is one of those agencies.

The following eight departments reside within the umbrella of the DOT. Each of these agencies are, or soon will be, in need of skilled employee support:

1. Federal Highway Administration (FHWA)
2. Federal Aviation Administration (FAA)
3. National Highway Traffic Safety Administration (NHTSA)
4. Federal Transit Administration (FTA)
5. Federal Railroad Administration (FRA)
6. Pipeline and Hazardous Materials Safety Administration (PHMSA)
7. Federal Motor Carrier Safety Administration (FMCSA)
8. Maritime Administration (MARAD)

On top of the thousands of vacancies that are becoming available, the BIL will cause another cascade of work to come flooding down the pike.

THE BIL

On November 15, 2021, President Joe Biden signed legislation for massive infrastructure spending: the $1 trillion Infrastructure Investment and Jobs Act—commonly referred to as the Bipartisan Infrastructure Law (BIL)—which has allocated funds to federal, state, and local governments to begin the task of repairing America's roadways.

According to a White House fact sheet, "This Bipartisan Infrastructure Deal will rebuild America's roads, bridges and rails, expand access to clean drinking water, ensure every American has access to high-speed internet, tackle the climate crisis, advance environmental justice, and invest in communities that have too often been left behind. The legislation will help ease inflationary pressures and strengthen supply chains by making long overdue improvements for our nation's ports, airports, rail, and roads. It will drive the creation of good-paying union jobs and grow the economy sustainably and equitably so that everyone gets ahead for decades to come. Combined with the President's Build Back Framework, it will add on average 1.5 million jobs per year for the next 10 years."[18]

The law provides for a mammoth investment of $350 billion in highway and roadway programs across the United States.[19] The sheer enormity of the roadway improvements and number of businesses that can prosper with this funding is difficult for even the most seasoned highway infrastructure professionals to grasp. The law allows local public entities and authorities—like state and local governments, metropolitan planning organizations (MPOs), and tribes—to apply directly for funding for community road improvement projects.

Sadly, US roadways have not been seriously invested in since 1957, at the end of Dwight D. Eisenhower's presidency. Let's take a short look back to learn some relevant history of the government's transportation objectives.

---

18   The White House, "Fact Sheet: The Bipartisan Infrastructure Deal," November 6, 2021, accessed September 12, 2023, https://www.whitehouse.gov/briefing-room/statements-releases/2021/11/06/fact-sheet-the-bipartisan-infrastructure-deal/.

19   US Department of Transportation Federal Highway Administration, Bipartisan Infrastructure Law, Funding, accessed September 12, 2023, https://www.fhwa.dot.gov/bipartisan-infrastructure-law/funding.cfm#:~:text=The%20Infrastructure%20Investment%20and%20Jobs,fiscal%20years%202022%20through%202026).

THE WORKS PROGRESS ADMINISTRATION (WPA)

"The Works Progress Administration (WPA) was an ambitious employment-and-infrastructure program created by President Franklin Roosevelt in 1935, during the bleakest days of the Great Depression. Over its eight years of existence, the WPA put roughly 8.5 million Americans to work building schools, hospitals, roads, and other public works. Perhaps best known for its public works projects, the WPA also sponsored projects in the arts—the agency employed tens of thousands of actors, musicians, writers, and other artists."[20]

This brings about a key point to the book: during Eisenhower's infrastructure build, many citizens of the time questioned why the government would hire artists and musicians, even at low wages, for roadway construction. This principle demonstrates that every walk of life is needed in this industry. The artist who designs and paints the murals on the subway wall has a place, just as the worker who lays the foundation of the structure does.

When the WPA was dissolved on June 30, 1943, the summary provided to Congress included the following remarks:

"The WPA built or improved 651,000 miles of roads, 19,700 miles of water mains, and 500 water treatment plants. Workers built 24,000 miles of sidewalks; 12,800 playgrounds; 24,000 miles of storm and sewer lines; 1,200 airport buildings; 226 hospitals; more than 5,900 schools, and more than two million privies."[21]

In 1956, twenty-two years after the establishment of the WPA and fourteen years after it ended, President Eisenhower signed the

---

20 History.com editors, "Works Progress Administration (WPA)," History.com, updated September 21, 2022, https://www.history.com/topics/great-depression/works-progress-administration.

21 Catherine Winter, "A Bridge to Somewhere," (WPA) Works Project Administration, accessed September 12, 2023, American Roadworks, https://americanradioworks.publicradio.org/features/infrastructure/b1.html.

National Interstate and Defense Highways Act. Eisenhower's push for interstate highways was a continuation of Roosevelt's 1938 Federal-Aid Highway Act. Any action on that study was delayed by World War II, but it was revisited by President Franklin Delano Roosevelt in his last full year in office. The Federal-Aid Highway Act of 1944 was the first initiative to build a formal interstate system, beginning with forty thousand miles of roads.

When Eisenhower assumed the presidency on January 20, 1953, he found the 1944 act had barely met 20 percent of the original goal of forty thousand miles. As an experienced soldier and leader of the war efforts, he understood the value of maintained transportation systems should the United States experience an invasion. Though he experienced some political snags before passing the law, ultimately an additional thousand miles were added to the original goal, and $25 billion in funds were allocated to build the US Interstate Highway System. The project broke ground in 1957 and went on for the next twelve years. A building boom had commenced.[22]

Certainly, the WPA and former leaders saw impressive achievements; however, they were accomplished with a devastating ulterior motive: our infrastructure itself was purposely designed to separate people into groups and keep them divided. We'll see in the following chapter just how this was achieved.

---

22   History.com editors, "The Interstate Highway System," History.com, updated June 7, 2019, History.com, accessed September 12, 2023, https://www.history.com/topics/us-states/interstate-highway-system.

2

DIVIDING HIGHWAYS—
DIVIDING SOCIETIES

When we talk about infrastructure being purposely designed to separate groups of people, we are in fact talking about physical barriers. The masses of concrete used to build the US highway system were strategically placed in order to contain minority populations and stymie their independent growth. How do we know this to be fact?

We can't even begin this conversation without mentioning Robert Moses, a powerful urban planner in New York post–World War II. Mr. Moses oversaw all public works projects in the state, and he is credited for designing a vast majority of bridges, tunnels, roadways, parks, and housing projects. His philosophy was to build highways directly through black neighborhoods in a plan to eliminate the "slums."

Farrell Evans, with History.com, explains,

"Moses, who was also the chairman of the New York City Slum Clearance Committee, said that the highway construction must 'go right through cities and not around them.' Two of the city's main arteries he created, the Cross-Bronx and Brooklyn-Queens Expressways, did just that, cutting through the heart of the Bronx and Red Hook neighborhoods.[23]

At the same time, the civil rights movement was becoming prominent. Congress and the federal courts made some progress by outlawing restrictive housing covenants and redlining, which essentially banned black homeowners from buying in white neighborhoods. Take a look at some of the language that was customary for the times:

The National Association of Real Estate Boards' *Ethics Handbook* published in 1924, contains this passage: "Article 84: A realtor should never be instrumental in introducing into a neighborhood a character of property or occupancy, members of any race or nationality, or any individuals whose presence will clearly be detrimental to property values in that neighborhood."[24]

A 1926 deed for a Syracuse property in the neighborhood of Scottholm says this: "Said lot shall not during the aforesaid period be occupied by or conveyed to negroes as owners or tenants."[25]

The very idea of integrated neighborhoods apparently sent panic racing through the establishment. A solution was devised that ended up being more costly than the millions of dollars invested into it. They would reinforce the segregated communities they wanted by literally dividing white neighborhoods from black, Hispanic, or Asian ones with massive road systems.

---

23   Farrell Evans, "The Interstate Highway System," History.com, updated June 7, 2019, History.com, accessed September 12, 2023, https://www.history.com/topics/us-states/interstate-highway-system.

24   Greta Kaul, "With covenants, Racism Was Written into Minneapolis Housing," Minnpost.com, February 22, 2019, https://www.minnpost.com/metro/2019/02/with-covenants-racism-was-written-into-minneapolis-housing-the-scars-are-still-visible.

25   Matt Mulcahy, "The Map: Urban Renewal and the 'Removal of Blacks' from the Center of Syracuse, CNY Central, March 23, 2021, accessed September 12, 2023, https://cnycentral.com/news/the-map-segregated-syracuse/the-map-urban-renewal-and-the-removal-of-blacks-from-the-center-of-syracuse.

As Farrell Evans explains, "According to estimates from the US Department of Transportation, more than 475,000 households and more than a million people were displaced nationwide because of the federal roadway construction. Hulking highways cut through neighborhoods, darkened, and disrupted the pedestrian landscape, worsened air quality, and torpedoed property values. Communities lost churches, green space, and whole swaths of homes. They lost small businesses that provided jobs and kept money circulating locally— crucial middle-class footholds in areas already struggling from racist zoning policies, disinvestment, and white flight."[26]

Unfortunately, Mr. Moses's impact on federal highway policy was adopted throughout the entire nation. Policymakers everywhere used our highway systems to deliberately break apart whole communities of black and poor residents. Any successful businesses or unified groups, such as church congregations or local coffee shops, were effectively split by concrete ramps, overpasses, and multilane highways. When it was done, America had a new infrastructure, a new landscape, and new immovable boundaries that contained disadvantaged groups of people. Worse, funding for city and state projects was often approved based on a municipality's plan to contain their black population.

As proof of this result, let's take a look at three cities where minority communities were decimated by the highway system.

---

26   Farrell Evans, "How Interstate Highways Gutted Communities—and Reinforced Segregation," History.com, October 20, 2021, https://www.history.com/news/interstate-highway-system-infrastructure-construction-segregation.

ORLANDO, FLORIDA, GRIFFIN PARK

Griffin Park. (Photo from HuffPost. © 2018 BuzzFeed Inc. All rights reserved. Used under license.)[27]

"Griffin Park is surrounded by two major highways that are used by hundreds of thousands of cars heading in and out of Orlando, Florida, every day. Whatever trees once buffered the noxious fumes and the roar of cars on all sides have been cut down. From above, you see a grid of apartment buildings encircled menacingly within a loop of the interchange, as if inside a noose.

"The pollution in Griffin Park and its low-income Parramore neighborhood is violence of a kind Americans tend to ignore. But it is as deliberate and as politically determined as any more recognizable act of racial violence. What happened to Griffin Park was the sum of a series of choices made over the course of a century, the effect of which was to transmute formal segregation into the very air certain people breathe."[28]

---

27   Julia Craven, "Even Breathing Is a Risk in One of Orlando's Poorest Neighborhoods," Huffington Post, January 23, 2018, https://www.huffpost.com/entry/florida-poor-black-neighborhood-air-pollution_n_5a663a67e4b0e5 630072746e.

28   Chris McGonigal, HuffPost, accessed September 12, 2023, https://www.huffpost.com/entry/ florida-poor-black-neighborhood-air-pollution_n_5a663a67e4b0e5630072746e.

ST. PAUL, MINNESOTA

Interstate I-94, downtown St. Paul, circa 1967. (Photo courtesy of Matt Reicher.)[29]

"The historic Rondo neighborhood once spanned much of what is now Summit Hill and ran north to University Avenue. By 1950, more than 80 percent of St. Paul's African American population lived there, leaving Rondo almost completely independent from the white neighborhoods surrounding it. But as the neighborhood thrived, Minnesota legislators began plans for a new highway system. In 1956, when the Federal Aid Highway Act was passed, St. Paul officials felt the pressure to start building. Concerns voiced by residents and city planners were almost entirely ignored as construction began on what would in 1968 become Interstate 94. Any debate over possible routes centered mostly on the concerns of (white) business owners, in their efforts to boost stagnating sales.

"When the route through the historic Rondo neighborhood was finalized, the city demanded that residents sell their homes to the city for dirt-cheap prices—often only a fraction of the actual property value. People that refused to vacate their homes and businesses were met by police with sledgehammers—destroying walls, smashing windows, and even tearing apart the plumbing. A lush and vibrant neighborhood was effectively sliced in half, displacing nearly 600 families,

---

29   Matt Reicher, "The Birth of a Metro Highway (Interstate 94)," Streets.mn, September 6, 2021, https://streets.mn/2013/09/10/the-birth-of-a-metro-highway-interstate-94/.

300 businesses and forcing thousands of African-Americans to seek alternative housing in a highly segregated city.

"It wasn't just physical—it ripped a culture; it ripped who we were ... It was an indiscriminate act that said this community doesn't matter ..."

—Minnesota governor Tim Walz[30]

SYRACUSE, NEW YORK

Construction of Interstate 81 through Syracuse, circa 1965. (Photo courtesy of The Onondaga Historical Association.)[31]

"In the 1960s, I-81 plowed through a historically black neighborhood in Syracuse, displacing hundreds ... Just south of downtown Syracuse in upstate New York, a stretch of highway has long divided surrounding neighborhoods.

---

30   Emma Nelson, "From Ashes to Asphalt: St. Paul's Systematic Destruction of Black Neighborhoods," Medium.com, March 1, 2017, accessed September 12, 2023, https://medium.com/@enelson009/from-ashes-to-asphalt-st-pauls-systematic-destruction-of-black-neighborhoods-54ea9c0c25f.

31   "The New York Highway That Racism Built: 'It Does Nothing but Pollute,'" The Guardian, May 21, 2021, https://www.theguardian.com/us-news/2021/may/21/syracuse-new-york-highway-i81-viaduct-biden.

"On the east side are large buildings where university students live, well-maintained green spaces, and a wall that blocks the highway from view. On the west side is a predominantly low-income and disinvested black neighborhood where the pollution from the highway exacerbates many residents' existing health conditions.

"For years, New York state officials have known that the ageing I-81 viaduct has needed to be radically redeveloped. Most residents and public officials agree that it must be rethought for safety, economic, and public health reasons. However, for a neighborhood that has long been disenfranchised, tearing down the highway also means repairing the legacy of injustice done to their community."[32]

• • •

Whole populations were blatantly contained with cement walls to reinforce segregation. These walls were also symbolic of another truth, unspoken yet loud in the ears of these disadvantaged residents: jobs that could pull them out of the cement blocks they'd been assigned to, such as high-paying DOT jobs, were not open to them. If they tried to apply, their applications were simply rejected.

As an African American woman in a leadership position within this industry, I have personally experienced some of the invisible barriers and lack of business opportunities minorities face every day. More often than not, we are counted out of the equation because of skin color, gender, age, sexual orientation, immigration status, or other personal biases.

In the real world, a "minority person" is often judged synonymously with "subordinate person." This being the case, my definition of *minority* is any group of people labeled as being subordinate to another dominant or specific group. So if we find ourselves in this category, what should we do about it? Let me answer that by

---

32   Rachel Ramirez, "The New York Highway That Racism Built: 'It Does Nothing but Pollute,'" *The Guardian,* May 21, 2021, accessed September 12, 2023, https://www.theguardian.com/us-news/2021/may/21/ syracuse-new-york-highway-i81-viaduct-biden.

conveying a true story about a poor immigrant who changed the transportation landscape.

In 1848, at the age of twelve, Andrew Carnegie immigrated with his family to America from Dunfermline, Scotland. Far from being an easy transition, the young boy witnessed his father begging for work, trying to make enough to support his family. He once wrote: "I began to learn what poverty meant." Andrew would later write, "It was burnt into my heart then that my father had to beg for work. *And then and there, came the resolve that I would cure that when I got to be a man.*"

Let's stop right there for just a moment. At this point, Andrew Carnegie, an impressionable young man, could have gone one of two ways: he could have gotten angry at the injustices his family endured, let himself fall into depression and/or unhealthy lifestyles, and done absolutely nothing to ensure change for himself and future generations. *Or* he could have looked the situation square in the eye, moved forward despite anyone else's opinion, and he could have radically made change that would positively affect the future. He chose to be the change.

Where did this type of independent—almost rebellious—way of thinking get Andrew Carnegie? He was a catalyst in America's participation in the Industrial Revolution, as he produced the steel to make machinery and transportation possible throughout the nation.

He was nicknamed "the Master of Steel," as he'd adopted Sir Henry Bessemer's new manufacturing process, which allowed steel to be made from iron more efficiently and quickly. This lowered the cost for steel, expanding the market. The steel produced worked extremely well for railways.

> "The Bessemer process was the first inexpensive industrial process for the mass production of steel from molten pig iron before the development of the open hearth furnace. The key principle is removal of impurities from the iron by

oxidation with air being blown through the molten iron. The oxidation also raises the temperature of the iron mass and keeps it molten."[33]

Andrew Carnegie's company, Keystone and Union Iron Works, also served as the general contractor to the St. Louis Bridge Company, which supplied the steel for the very first steel bridge, the Eads Bridge. The Eads Bridge is a combined road and railway bridge over the Mississippi River that connects the cities of St. Louis, Missouri, and East St. Louis, Illinois.[34]

> "The St. Louis bridge vaulted 1,500 feet across the Mississippi with a main span of 515 feet. It was, in the words of the great architect Louis Sullivan, 'spectacular and architectonic.'"[35]

Just as Andrew Carnegie did during the Industrial Revolution, we have an opportunity to change our world and form a better future. Like many minority groups, he watched the struggles of his father. Instead of begging for an opportunity, *he capitalized by solving a problem the country was having.*

I've identified five specific challenges that are routinely problematic in the minority woman's career life. Acknowledge and address these barriers as you move forward. Like Andrew Carnegie, take what you see and change the world. Create a more inclusive and diverse industry, where everyone has an equal opportunity to succeed.

---

33 Wikipedia, s.v. "Bessemer Process," accessed September 12, 2023, https://en.wikipedia.org/wiki/Bessemer_process.

34 Wikipedia, s.v. "Eads Bridge," accessed September 12, 2023, https://en.wikipedia.org/wiki/Eads_Bridge.

35 William Dietrich II, "Andrew Carnegie: The Black and the White," Pittsburgh Quarterly, summer 2007, accessed September 12, 2023, https://pittsburghquarterly.com/articles/andrew-carnegie-black-white/.

LACK OF REPRESENTATION

Women from minority backgrounds often struggle to find role models and mentors who can guide them in their industry. We've already discussed how inequality has been present in the transportation industry. For well over a century, it's been dominated by white males. While women and minority groups have fought valiantly against discrimination, underrepresentation continues to plague minorities, especially black women.

Mentorship programs, formal or informal, can do wonders for fostering growth and providing a sense of representation, or someone "having your back," by connecting new talent with seasoned professionals who can serve as role models as they begin building their careers. Establishing initiatives, community forums, social groups, and clubs where members can confidentially talk about their experiences and feelings creates "representation."

When I first envisioned Highway Infrastructure Courses Online, my intention was to assist in leveling the playing field of the underrepresented. The website and courses currently under development will be geared toward providing quality training and positive career connections. We're developing specific resource groups to help students find mentors and learn about career opportunities. Please scan the QR code below to learn more about being a part of our tribe.

BIAS AND DISCRIMINATION

Implicit biases and stereotypes can lead to prejudice and discrimination against minority women. For instance, some stereotypes paint the minority female as less intelligent than nonminority counterparts if she has an accent. This is a blatant untruth, yet mindsets such as this make it harder for these women to be taken seriously and limit them from being given equal opportunities for advancement.

This is a fact that minority women must manage as they enter their daily businesses. It is important to understand who you are as an individual and to decide that if you are not accepted or taken seriously in a room, you'll go right ahead and find another room. It is paramount to avoid begging for inclusion at the table of another; it is imperative to acquire the wisdom and proficiency to construct your own table of utmost autonomy, wherein the privilege of determining the people to be seated alongside you rests solely with you.

NETWORKING CHALLENGES

Building professional networks is crucial for career growth, but minority women may face challenges in accessing these networks due to exclusion, limited connections, lack of inclusion in industry events, or a prevailing sense of discomfort in an already-uncomfortable environment.

These networking opportunities are crucial, especially in any career field. You may have to step out of your comfort zone and take risks to become involved. Again, build your own networking groups if no existing groups are established. I've found that a world of opportunities, and people who want to support others, exists.

UNEQUAL ACCESS TO RESOURCES

Limited access to financial resources, educational opportunities, internships, and training programs can put minority women at a disadvantage. The fear of confirming negative stereotypes about one's race or gender can create additional pressure and stress, affecting performance and confidence levels. There is a provable pattern of financial rejection that minority women must often plan for and overcome. Let me back that statement up with a true story:

With my first entrepreneurial endeavor, I started with minimal cash and one contract worth $6,000 from the city of Grand Prairie, Texas. As I progressed throughout the years, I went to my bank for a loan so I could hire an employee, and unfortunately, I was denied, even though I could prove steady income and a credit score of 710.

Fast-forward to a few years later: After having started multiple business ventures, this time I had white male partners. The same type of situation came up, except this time, lo and behold, the loan was approved.

This type of treatment pervades the professional lives of minority women. Instead of acceptance based on merit, we see acceptance based on biases or unfair judgment.

WORKPLACE CULTURE BIASES / LACK OF SUPPORT

The prevailing culture and bias within organizations can create a hostile or unwelcoming environment for minority women, making it harder for them to thrive and reach leadership positions. Similarly, minority women may face a lack of support from colleagues, superiors, or even family members who may not fully understand or believe in their ambitions and abilities.

Whether or not this continues is really up to each individual experiencing it. Again, I encourage you to locate professional groups, or develop your own, where members can find support for their professional lives. If you find your group needing guidance, follow my "SIS" model: let's support, inspire, and embody self-care.

Support

Benjamin Disraeli is quoted as saying, "The greatest good you can do for another is not just to share your riches, but to reveal to him his own."[36] This is so true. Women, regardless of culture, need to have at least a small group where harsh criticism is banned, wins are celebrated, and support is both emotional and practical.

Inspire

Inspiring one another is key. Inspiration is something that believes in itself and leads us to the best of ourselves. Women need inspiration every single day as they not only face professional challenges but, oftentimes, personal and physical ones as well. So cultivate the minds of those coming under you; embrace the wisdom of those who came before you; and inspire those around you by being an inspirational example.

Self-Care

An essential and often overlooked priority of any group focused on supporting their fellow female professionals is self-care. Whether faced with outright workplace biases or lack of professional support on any other level, finding your own strengths will help identify where to find your support. I've observed three keys to professional women finding *the right support* for their specific needs:

36   Benjamin Disraeli, "Quotes," Goodreads, accessed September 12, 2023, https://www.goodreads.com/quotes/17581-the-greatest-good-you-can-do-for-another-is-not.

You can join my Facebook Group: Women of Purpose Business Forum
for information and resource information:
https://www.facebook.com/share/g/8aWn4xzhe6y15mu6/?mibextid=K35XfP

1. **"Reflect, set, and say no."** This is a great motto to live by when you're managing a busy professional life. The strategy is to face most situations with this pattern of thinking:

   □ Reflect on your values, beliefs, or sense of purpose and find ways to incorporate them into your daily life. Engage in practices that nurture your spiritual well-being, whether it's through prayer, meditation, journaling, or seeking guidance from a spiritual advisor.

   □ Set boundaries to help prevent burnout and allow you to focus on activities that bring you joy and fulfillment.

   □ Say no! Undoubtedly, for most of us, setting boundaries will include learning to say no. Most professional women are juggling homelife, personal life, and work life at an ever-frantic pace. Declining a social invitation, or even new clients, may be necessary if you don't have the mental or emotional energy to carry through and recover from the experience.

Professional therapy, or a combination of therapy and mentorship, can be tremendously helpful in identifying and accepting our own skill levels and the mindsets that hold us back.

2. **Establish health routines.** We're not talking about rigorous, time-consuming health routines here, but practical ways to fit health into your busy schedule. Focus on three simple-yet-life-altering points of health: diet, exercise, and sleep.

▫ **Diet:** In conjunction with intermittent fasting, incorporating a variety of fruits, vegetables, whole grains, lean proteins, and healthy fats into your meals has proven benefits to your health. Not only will this type of reasonable eating help maintain a healthy weight, but these foods also can be prepared as plain or as extravagantly as you'd like.

Staying hydrated by drinking enough water throughout the day is one of those simple health changes that can instantly alter the way you feel. When you're feeling that midday fatigue and you must carry on, try drinking a full glass of water. In no time, you'll likely feel energy and focus coming back into balance.

▫ **Exercise:** Occupied professionals can get lost in time while working, and before they know it, they haven't moved for eight solid hours. Physical activity not only benefits your physical health but also releases endorphins that boost your mood. This, in turn, enhances your work performance!

Engage in regular exercise or physical activities that you enjoy. Try walking, running, joining a fitness class, practicing yoga, dancing, or playing a sport both for the physical benefits and to decompress your mind. For ideas, see my Amazon storefront, https://www.amazon.com/shop/estherfranklin_official.

▫ **Sufficient sleep:** A regular sleep schedule and quality sleep each night ensures your body and mind are well rested. It cannot be overstated that lack of sleep will adversely affect your brainpower and memory, the way you look, and how well you feel physically.

The long-standing mantra of "early to bed; early to rise" is actually a proven strategy for optimum sleep health. Aim for seven to eight hours of quality sleep every night; it's crucial for your well-being. Your wealth is your health.

3. **Invest in yourself.** "Investing in yourself" is oftentimes over-looked when ambitious people tackle their careers. Young professionals, especially, can see this type of investment as "wasting time." You couldn't be more wrong! Engaging in activities that bring you joy and help you relax, such as partaking in hobbies, listening to music, or watching your favorite movies or TV shows, has a renewing effect on us.

Anyone who has experienced burnout will tell you that investing in yourself—taking those breaks to fully engage in activities that relieve your mind and bring fulfillment to the soul—should be seen as a business strategy, regularly scheduled into your calendar. Directing your mind and body toward interests outside of work often revitalizes your work performance.

• • •

Under the provisions of the BIL, and with the actions provided throughout this chapter, the infrastructure that divided and enclosed whole people groups can be redesigned by those it oppressed. Rose Morrison lists seven ways the BIL supports disadvantaged groups in her article "How the Infrastructure Bill Could Expand Opportunities to Women and Minorities":

1. Supports women and minorities in the construction industry
2. Increases the number of trade positions
3. Supports free universal preschool
4. Reforms the immigration system
5. Supports technological advancements
6. Expands career-building tools (high-speed internet)
7. Takes necessary harassment-improvement measures[37]

Women and minority groups, you are especially *encouraged* to take leadership positions. *This is your time!*

---

37   Rose Morrison, "How the Infrastructure Bill Could Expand Opportunities to Women and Minorities," Fieldwire by Hilti, March 28, 2022, https://www.fieldwire.com/blog/infrastructure-bill-opportunities-for-women-and-minorities.

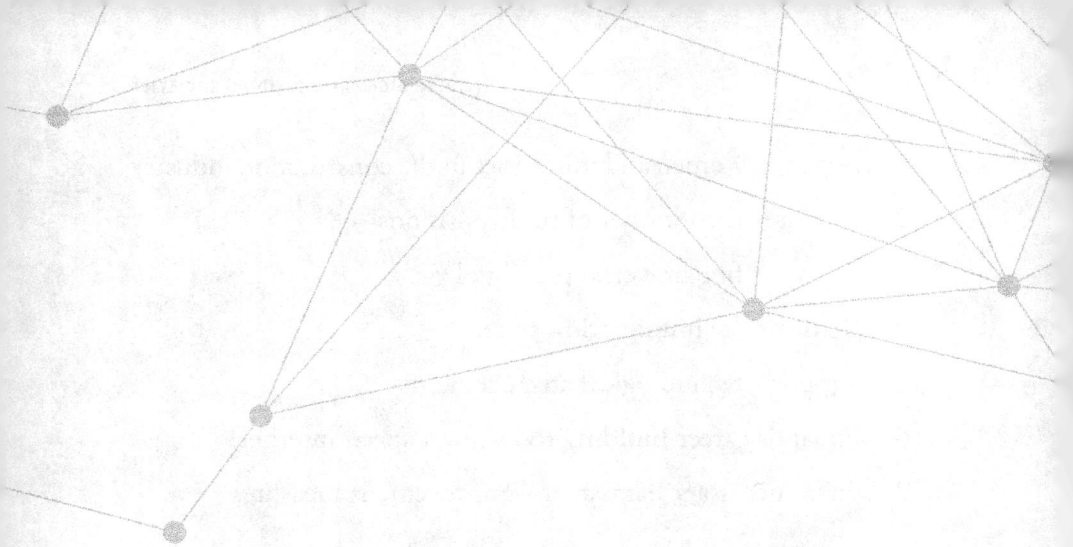

3

WOMEN AND MINORITIES WELCOME

As the transportation industry explodes, it happens to be in the historic position of having predominantly female leaders. It's difficult to convey how much of a "shattering of the glass ceiling" this is. *Bloomberg Law* sums it up well in their article "Women Breaking Ground in Shaping Infrastructure Plans":

"For the first time in history, staffs of both the House Transportation and Infrastructure Committee and the Senate Environment and Public Works Committee are led by women, and the subcommittees overseeing transportation programs also have female staff directors. The rise of senior women staffers mirrors the increasing number of female lawmakers, with 142 women in voting positions now—or about 26 percent of seats—up from just 26 women in the late 1980s."[38]

As I mentioned, women—especially women with minority backgrounds—*this is your time.* There is a shift happening; minority groups are becoming the majority, and minority women are taking lead roles throughout the transition. There are regulations, rules, and laws in place to protect women and minority groups from discrimination and support minority-owned endeavors. Disenfranchised groups are encouraged to take on strategic positions in the restructuring-roads process. *Companies who want the infrastructure contracts resulting from the BIL must allocate a certain percentage of the project to these groups.*

Four specific programs have been mandated to promote and support fairness to minority business entities, especially in the construction and transportation fields: Affirmative Action, SBA 8(a), Disadvantaged Business Enterprises, and the Reconnecting Communities Program. Let's take a close look at each one.

MINORITY-FOCUSED FEDERAL PROGRAM #1: AFFIRMATIVE ACTION

"Affirmative action is defined by OFCCP regulations as the obligation on the part of the contractor to take action to ensure that applicants are employed, and employees are treated during employment, without regard to their race, color, religion, sex, sexual orientation, gender

---

38 Nancy Ognanovich and Lillianna Byington, "Women Break Ground in Shaping Infrastructure Plans in Congress," *Bloomberg Law*, May 6, 2021, https://news.bloomberglaw.com/social-justice/women-break-ground-in-shaping-infrastructure-plans-in-congress.

identity, national origin, disability, or status as a protected veteran (dol.gov)."[39]

Affirmative action was established in the 1960s during the Lyndon B. Johnson administration. Its purpose was to ensure that minority groups had a fair chance at receiving good-paying employment. This was the first real step toward equalizing opportunities for job candidates with minority race or ethnic backgrounds.

MINORITY-FOCUSED FEDERAL PROGRAM #2: SBA 8(A)

Another program that was developed to assist women and minority groups in their success is SBA 8(a). Business owners with disadvantaged backgrounds who are interested in expanding into the federal space can obtain valuable business assistance within the program. Business owners must be in business for at least two years to qualify, and acceptance into the program does not guarantee any contracts. However, take a look at the astounding benefits the US Small Business Administration lists for certified firms in the 8(a) program:

- "Efficiently compete and receive set-aside and sole-source contracts
- Pursue opportunity for mentorship from experienced and technically capable firms through the SBA Mentor-Protégé program
- Pursue joint ventures with established businesses to increase capacity

---

39 US Department of Labor, Office of Federal Contract Compliance Programs, Affirmative Action, "Frequently Asked Questions," updated on October 28, 2022, https://www.dol.gov/agencies/ofccp/faqs/AAFAQs#:~: text=Affirmative percent20action percent20is percent20defined percent20by, percent2C percent20national percent20origin percent2C percent20disability percent2C percent20or.

- Qualify to receive federal surplus property on a priority basis
- Receive one-on-one business development assistance for their nine-year term from dedicated Business Opportunity Specialists focused on helping firms grow and accomplish their business objectives
- Connect with procurement and compliance experts who understand regulations in the context of business growth, finance, and government contracting
- Receive free training from SBA's 7(j) Management & Technical Assistance program."[40]

When a company has an 8(a) certification, it means their business is eligible to compete for the designated contracts within the program. Because the federal government has a current employment need plus a mandate to hire equitably, this is an unprecedented opportunity for historically underrepresented entrepreneurs to win lucrative contracts in the field of transportation.

To apply for the 8(a) program, follow these steps:

1. Identify your primary NAICS code(s).
2. Register your business in the System for Award Management (SAM).
3. Apply for 8(a) certification.

Visit the Knowledge Base at www.sba.gov to find helpful resources, including the application guide, to assist with gathering

---

40 "US Small Business Administration, 8(a) Business Development Program," accessed September 12, 2023, https://www.sba.gov/federal-contracting/contracting-assistance-programs/8a-business-development-program.

necessary documentation as well as completing and submitting the application.[41]

If you are interested in registration, the Small Business Administration has a streamlined process to assist you: https://www.sba.gov/federal-contractingcontracting-assistance-programs/8a-business-development-program.

MINORITY-FOCUSED FEDERAL PROGRAM #3: DISADVANTAGED BUSINESS ENTERPRISES

There are two parts involved in any project supported by federal funding: the project itself and the companies who make each piece of it a reality. Let's say your city legislators have decided to extend Main Street in your area and connect it to the entrance ramp of the interstate. To make that happen, they'll apply for federal funding to complete the extended road. In order to receive those funds, the municipality, transportation authority, or institution overseeing the project must allocate a certain amount of work to firms listed on an approved Disadvantaged Business Enterprise (DBE) list.

The Small Business Administration defines DBEs as "for-profit small business concerns where socially and economically disadvantaged individuals own at least a 51 percent interest and also control management and daily business operations."[42]

"In 1983, Congress enacted the first Disadvantaged Business Enterprise (DBE) statutory provision. This provision required the

41   US Small Business Administration, "8(a) Business Development Program," accessed September 12, 2023, https://www.sba.gov/federal-contracting/contracting-assistance-programs/8a-business-development-program.

42   US Department of Transportation, "Definition of a Disadvantaged Business Enterprise," accessed September 12, 2023, https://www.transportation.gov/civil-rights/disadvantaged-business-enterprise/definition-disadvantaged-business-enterprise.

Department to ensure that at least 10 percent of the funds authorized for the highway and transit Federal financial assistance programs be expended with DBEs."[43]

On the flip side of this project are the actual construction, blacktop, heavy-machine companies, etc., that will install that Main Street extension. If they are registered DBE companies, they have a chance to be prioritized for that project.

Disadvantaged Business Enterprises (DBEs) (also Minority and Women-Owned Business Enterprises; M/WBEs) are offered specific and tremendous opportunity in the infrastructure rebuild. According to the Code of Federal Regulations, the DBE Program has eight objectives that recipients of DOT funds must develop and implement. As you read these benefits, keep in mind that companies contributing to the infrastructure rebuild will be *required* to use DBEs.

1. "To ensure nondiscrimination in the award and administration of DOT-assisted contracts in the Department's highway, transit, and airport financial assistance programs;

2. To create a level playing field on which DBEs can compete fairly for DOT-assisted contracts;

3. To ensure that the Department's DBE program is narrowly tailored in accordance with applicable law;

4. To ensure that only firms that fully meet this part's eligibility standards are permitted to participate as DBEs;

5. To help remove barriers to the participation of DBEs in DOT-assisted contracts;

6. To promote the use of DBEs in all types of federally-assisted contracts and procurement activities conducted by recipients;

---

43   US Department of Transportation, "History of the DOT DBE Program," accessed September 12, 2023, https://www.transportation.gov/osdbu/disadvantaged-business-enterprise/history-dot-dbe-program.

7. To assist the development of firms that can compete success-fully in the marketplace outside the DBE program;

8. To provide appropriate flexibility to recipients of Federal financial assistance in establishing and providing opportunities for DBEs."[44]

To be clear, the DBE Program is federally funded; however, an organization must be certified as a DBE through the state where the organization is located. Be alert to your state because specific states may offer additional funding available through their own programs. The DBE Program has been renewed several times, and it's still active today, as the BIL itself is explicit in its investment into this program. DBEs and those who are already established are poised to *reap the business* as contracts roll out!

DBE Participation Goals

If a public works project is federally funded, there is a "DBE Participation Goal" that must be fulfilled. This means the company receiving federal financing for the job must include DBEs for 10 percent or more of the overall project. There must be a DBE firm participating in the contract, whether they serve as the prime contractor or serve as the subcontractor to the prime.

These goals ensure that a percentage of the project's contracting opportunities are reserved for these businesses. By participating in this program, DBEs and Women/Minority Business Enterprises (W/MBEs) can gain access to a larger pool of potential contracts and business opportunities.

---

44 Code of Federal Regulations, "A Point in Time eCFR System," National Archives, last amended March 6, 2023, accessed September 12, 2023, https://www.ecfr.gov/current/title-49/subtitle-A/part-26/subpart-A/section-26.1.

The inclusion of DBEs and W/MBEs in federally funded projects promotes economic equity, diversity, and increased competition in the marketplace. Smaller businesses and historically marginalized communities are *needed* to participate in infrastructure development and construction projects.

MINORITY-FOCUSED FEDERAL PROGRAM #4: RECONNECTING COMMUNITIES PROGRAM

US Transportation Secretary Pete Buttigieg announced on February 28, 2023, that forty-five projects, budgeted at $185 million in grants, were being awarded through the Reconnecting Communities Pilot Program.

According to the Department of Transportation, "The purpose of the RCP Program is to reconnect communities by removing, retrofitting, or mitigating transportation facilities like highways or rail lines that create barriers to community connectivity, including to mobility, access, or economic development. The program provides technical assistance and grant funding for planning and capital construction to address infrastructure barriers, reconnect communities, and improve people's lives."[45]

If you recall from chapter 2, we discussed how segregation had been deliberately built into our highway system. The Reconnecting Communities Program specifically counters this deliberate attempt to suppress minority groups. The program recognizes the communities that have been affected and provides funding for roads and/or bridges to be built that specifically connect those communities to vital

---

45   US Department of Transportation, "Reconnecting Communities—FAQs, RCP Grant Priorities," accessed September 12, 2023, https://www.transportation.gov/grants/reconnecting-communities/reconnecting-communities-faqs.

resources, such as job opportunities, medical services, and shopping centers. The following funding for the Reconnecting Communities Program Grants is published on the transportation.gov site:

Fiscal Year	2022	2023	2024	2025	2026	5-Year Total
Planning & Technical Assistance	$50M	$50M	$50M	$50M	$50M	$250M
Capital Construction	$145M	$148M	$150M	$152M	$155M	$750M
Total Authorized Amount	$195M	$198M	$200M	$202M	$205M	$1,000M[46]

To be clear, businesses don't qualify for the grant; only the state, tribal governments, and municipalities do. Businesses will bid on the projects funded by the grant. The grant funds the planning activities that will be used to determine the cost associated with the removal of any barriers their predecessors strategically imposed.

The Federal Highway Transportation Authority and the US DOT are the agencies authorizing funding. There will be millions of dollars in projects coming in the next several years, as you can see from the chart above. Following are just twelve of the forty-five grants awarded in FY22.

I encourage you to scan the QR code below to identify if your state has received funding, so you may join in on the rebuild of your community. Get in where you fit in!

---

46   US Department of Transportation, "Reconnecting Communities Pilot Program," accessed September 12, 2023, https://www.transportation.gov/grants/reconnecting-communities/reconnecting-communities-fy22-award-fact-sheets.

Reconnecting Communities Pilot Program FY 2022 Award Fact Sheets

Grant Type	Project	Applicant	State	Amount
Capital	Shoreline Drive Gateway	City of Long Beach	California	$30,000,000
Capital	Uniting Neighborhoods & Infrastructure for Transportation Equity (UNITE): Ashley Drive	City of Tampa	Florida	$5,354,695
Capital	The City of Kalamazoo: Reconnecting Communities Pilot Project for Kalamazoo–Michigan Avenues	City of Kalamazoo	Michigan	$12,272,799
Capital	Bridging I-696: Connecting Oak Park	Michigan Dept. of Transportation	Michigan	$21,704,970
Capital	NJ TRANSIT's Long Branch Station Pedestrian Tunnel	New Jersey Transit Corporation	New Jersey	$13,215,036
Capital	NYS Route 33 (Kensington Expressway) Project	New York State DOT	New York	$55,597,500
Planning	Birmingham Transportation Capital Investment Plan	City of Birmingham	Alabama	$800,000
Planning	Reconnecting Fairview: Neighborhood Revitalization through Community-Led Highway Redesign	Anchorage Neighborhood Housing Services	Alaska	$537,660
Planning	Atravesando Comunidades: Tucson's Greenway and Bike/Ped Bridge Project	City of Tucson	Arizona	$900,000
Planning	Little Rock I-30 Deck Park Phase I Planning Study	City of Little Rock	Arkansas	$2,000,000
Planning	Vision 980 Study Phase 2 Feasibility Study	California Dept. of Transportation	California	$680,000
Planning	SR-710 Northern Stub Re-envisioning Project	City of Pasadena	California	$2,000,000

4

HIGH SCHOOL AND COLLEGE STUDENTS, A BRIGHT FUTURE AWAITS

Because I understand the hard work of "figuring it out" and going after education without direction, I clearly see that young people planning their career paths are in the most advantageous positions.

At the time of this writing, we've pretty much gotten through the COVID-19 pandemic. However, there were millions of high school students who were greatly confused and stifled because of the upheaval. Many were so thrown that they couldn't regain their footing. Working in the transportation industry could provide them a chance to skip right over that stress, move forward on their own terms, and start the rest of their lives with stability and good-paying jobs.

When I graduated from high school—and even when I graduated from college—I didn't know title and right-of-way work even existed. I can't recall even one classmate, either, who set out to work in this field.

It was virtually unknown. When I was exposed to it and realized the capabilities included good pay and an opportunity to be in business for myself, I became determined to figure things out. As a strategy, I worked backward, in a way, using three simple steps. I recommend using this method and streamlining the process as much as possible:

1. First, decide how much money you need/want to earn.

2. Look at job descriptions in detail and make lists of those that match your interests and skills.

3. Go after whatever education, licensing, or certification is needed to make your dream job your reality.

In my case, because there was no direction in the very beginning, the process took a lot of time. My sister used to call me a professional student and razz me about what I was going to do with all of my education. If I'd had someone to tell me about a field on the verge of needing my talents, I would have streamlined my efforts!

Students, you can avoid years of wandering before finally reaching your goals. Find the job that suits you best as you continue through the book, and learn more about these opportunities. As you conquer your path, keep these five tips in mind:

1. BE READY, WILLING, AND ABLE TO WORK HARD

In all fairness, some individuals just want a paycheck; they're interested in providing the essentials and maybe having a few extras; they are not so much interested in being part of change, especially when that change becomes uncomfortable.

The five careers needing immediate fulfillment that are high-lighted in this book are different. I see many who begin learning and then give up when the challenges get thick. Most positions will surely include days where you wonder how to accomplish different aspects of the job. You'll need to have the capability and willingness to figure it out. There may be times when things come across your desk that you really don't have expertise in. Take on an attitude of determination, *become* the expert, and learn what's needed to successfully close that project.

The careers we will highlight will be a part of making the future better for you, your children, and generations to come. You can and must face every task, every day, with purpose. A failure mentality will not succeed in this industry, no matter what your ethnic background is.

2. CHANGE YOUR MIND

If negative feelings arise because you or your loved ones personally experienced deliberate segregation, you're not alone. The question becomes, What are you going to do about it? Are you going to stew in your anger and pretend you're hurting others with it? Or are you going to grab ahold of the opportunities in front of you to become independent, a living part of the future?

I want to encourage you to focus any negativity into a positive *for yourself and your family*. Be one of the people to actually make a difference and say, "Let's build this bridge here to connect these communities for the good of the collective as opposed to being segregated with the intention of keeping people divided."

3. GET THE PROPER TRAINING

Some very lucrative positions in title and right-of-way work do not technically require college. If you're a critical thinker and you can comprehend and apply specific procedures, a high school diploma may be sufficient to begin training.

Now, before dropping all of your plans for a college education, think carefully. The current advancements in technology may require at least an associate's degree for trainers to take you seriously. Why is that? Let me explain by providing a true example.

I know of an office that was looking for an administrative assistant. One of the accountants had a daughter who had just graduated from high school, and everyone wanted to give her a chance. When interviews came along, however, it was clear that she did not know some basic office technology that was critical to run the company. She could text or search the web like a pro, but they needed familiarity with actual phone systems and computer programs, as their day-to-day business was quite complex. She could have eventually been brought up to speed, but there wasn't the time, and it wasn't cost-effective to provide that much training. Unfortunately, she was eliminated as a candidate.

Technology expands so rapidly these days that, in my opinion, college or technical training will only help you, even if it's not expressly required for the position that suits you.

4. STRATEGIZE OPEN-MINDEDLY

For college students, much confusion can be spared by knowing your path before matriculating. Current graduates with degrees are often-times stunned when they cannot find a job. Knowing that an industry is about to need your talents can help you steer your education into something rewarding and profitable. Applying your passion to the transportation industry is a sound path forward.

When you consider career options, don't forget the trades. No road will be built, bridge constructed, track laid, or project completed without heavy machine operators, welders, plumbers, and other tradespeople to physically "build" our new infrastructure. Skilled construction workers, like electricians, blacktop installers, and specialized-equipment operators, are crucial to the planning and long-term success of infrastructure projects.

The jobs that will be created promise to be lucrative, long lasting, and even multigenerational, as some projects will span ten, twenty, or thirty years, or more. The BIL offers anyone seeking a new career—including high school graduates, those finishing their two- or four-year degrees, or mature job seekers—unprecedented opportunities to get on the ground floor of the biggest infrastructure building boom this country has seen to date.

5. REMAIN INDEPENDENT

A cultural phenomenon of our times is a general idea that a "successful life" includes having someone else to support us, whether through government funds or by relying on parents or a spouse. It's difficult to reach those who truly believe this, but here is the reality: no one owes you anything, except you. You owe yourself your best life, with a fulfilling job and good money.

Young people—and especially young women of all races—please heed this warning: the minute you put your life in somebody else's financial care, they have the ability to dictate and limit your lifestyle. It's a position that our female ancestors fought courageously against. I urge you to work hard to remain independent, provide for yourself and your family, and reach for those paydays that keep you in control of your own destiny.

KEYS TO SUCCESS:
HARD WORK AND A GREAT MENTOR

After graduating from college, I started out as an intern making $5.15 an hour. Looking back on that now, I wonder what I was even thinking. This step was necessary, however, to get the experience I needed to move on to bigger and better things later. This is why you'll hear me emphasize the strategy of interning while you're a student, especially if you live at home. Those smaller interning wages are easier to stretch if you don't have to fully support yourself. If you time it properly, you can "intern while you learn" and come out of school ready to launch.

You'll find that many positions require knowledge of processes in addition to either high school or college. So how does one develop these skills? Without a doubt, having a mentor to support you as you learn your trade is the best way to get the experience needed.

My introduction into title and right-of-way work came about because my paralegal company is a certified DBE. Fred Brient, with Orion Land Service, happened to be looking for a minority business enterprise to team up with to fulfill his participation goal. He thought I'd be perfect for title work, and he introduced me to the field. Fred was a seasoned, sharp-witted veteran in title and right-of-way procedures. He was as tough and as driven as he had to be in order to get the job done, and he pushed me toward excellence. I'll be forever grateful for that. Experience is vital before going after contracts on your own.

The career paths we'll discuss each have differing degrees of mentoring requirements. Some careers literally require oversight before licensure can be achieved, and for others, mentorship is more of a recommendation. Either way, choose your mentors wisely! They not only provide oversight, but they help their mentees cultivate their craft.

5

THE REAL ESTATE SECTOR AND NON–REAL ESTATE SECTORS

THE REAL ESTATE CONNECTION

For the purposes of this book, we can break down the five careers needing immediate candidates by "real estate sector" and "non–real estate sector." Engineers and architects, as you'll see in great detail in the upcoming chapters, plan and build highway infrastructure. They focus their talents and resources on the structure being built. These career paths fall into the non–real estate sector. Why? Because engineers and architects are critical to a highway infrastructure project ever beginning, but they have nothing to do with the actual ownership of the land being built upon. Those details fall into what is known as the real estate sector.

Very simply, the real estate sector is made up of those careers that deal with the real estate needed for highway infrastructure projects. Roads and bridges cannot just be installed on property owned by private citizens. The land must be negotiated for and purchased according to established laws. Therefore, it must be appraised, and its ownership must be verified. The real estate sector careers deal with this type of business, and we will highlight these three:

1. Right-of-way experts (agents and professionals)
2. Appraisers
3. Title experts (abstractors and agents)

For both sectors, if ambitious people do not take hold of these careers, the consequences will be the same: project completions will quickly begin to lag, and the problem will only get more pronounced as newly approved projects commence.

REAL ESTATE SALES: EARN WHILE YOU LEARN

Notably, the career paths of right-of-way experts, appraisers, and title experts can all benefit from a start in real estate sales. I can't emphasize enough, that "earning while learning" is the best way to progress forward during any necessary training phases. Many of these careers take time before trainees are on their feet and earning the money they are after—there is no way around that. Life responsibilities and financial needs change as time passes, and real estate sales can offer a good financial living plus necessary industry connections and understanding of infrastructure projects.

The Five Careers Needing Immediate Fulfillment

As we've discussed, the five careers needing immediate fulfillment concerning the infrastructure rebuild are engineers, architects, right-of-way experts, appraisers, and title agents. For an authentic picture of each career, over the next five chapters, an industry expert in each of these fields will take the lead and present the chapter to you.

If you are qualified for one of these five positions, work awaits you now, so go ahead and pursue that career path. If one of these careers interests you, but you *do not* have the qualifications, I hope you'll take the necessary steps, whether through education or training, to grab ahold of your opportunity. A stable industry that pays well needs *you*.

So let's start with the first career on our list, engineers. What do they do? What kind of educational requirements are needed to become an engineer? Could you be a good candidate for this line of work? Let's find out.

PART 2

CHOOSING YOUR LANE: THE NON-REAL ESTATE SECTOR

ENGINEERS AND ARCHITECTS

6

ENGINEERS

Career Information

PROVIDED BY PROFESSIONAL ENGINEER
JARED M. GREEN, PE, D.GE, F.ASCE

Jared is originally from southwest Philadelphia, Pennsylvania. He graduated from Syracuse University's College of Engineering with a BS in Civil Engineering and later went on to attain his MS in Civil Engineering from the University of Illinois Urbana-Champaign. He began working in the NYC office of Langan Engineering and Environmental Services, Inc., over twenty years ago and is now a principal / vice president and geotechnical practice leader in the Philadelphia and Pittsburgh offices. Jared is a licensed professional engineer in fourteen states and has been inducted into the Academy of Geo-Professionals.

Jared leads teams that are responsible for projects throughout the mid-Atlantic region; he's a consultant and a team leader, and he enjoys mentoring young engineers and first-generation college students. As such, he's been instrumental in increasing the number of precollege students who are interested in **STEAM** (science, technology, engineering, arts, and mathematics) and is regarded by many as a collaborator and a problem solver at heart.

Jared resides in New Jersey with his wife and their three children.

About the Career

Definition of Engineer (Cambridge Dictionary): "A person whose job is to design or build machines, engines, or electrical equipment, or things such as roads, railways, or bridges, using scientific principles.[47]

National Average Salary (Bureau of Labor Statistics): $88,050 yearly; $42.33 per hour.

Current State of Engineering in Infrastructure: Needing immediate candidates.

Opportunities to Work for a Company? When we're talking about infrastructure projects, a lot of times, there are multiple companies involved, with multiple engineers collaborating. To be directly involved in the upcoming infrastructure reform, candidates will likely find those opportunities at large firms.

Opportunities for Self-Employment? Yes; however, project leaders will want to know who you've worked for and what you can do. In my experience, the most successful engineering entrepreneurs have been in the industry for ten to fifteen years. They've designed something that has been constructed, and they're familiar with processes and pitfalls. Business owners need their license, insurance, registration with a local jurisdiction, an address, a company structure, an accounting and invoicing system, etc. Those planning on opening their own firm in the future would be wise to minor in entrepreneurship or something similar.

THE CAREER OF ENGINEERING

When a building, bridge, tunnel, box culvert, etc. is being built, engineers get involved in the beginning planning stages. We work very closely with the architects who are designing the actual structure involved in the endeavor. For example, let's say the Syracuse DOT is building a bridge, and they specifically want it to span over an existing highway. To accommodate this, the architect decides that an arch bridge would serve the purpose very well. Before this design takes

---

47   *Cambridge Dictionary*, s.v. "engineer (*n.*)," accessed September 12, 2023, https://dictionary.cambridge.org/dictionary/english/engineer.

place, the architect and engineering team will collaborate to be sure the design elements are achievable.

Notably, not just one engineer will work on this project. Specifically for a bridge build, a team will be assigned that includes hundreds of people who are responsible for the actual design and thousands of people who are responsible for the design, construction, operation, and financing.

TYPES OF ENGINEERS IN AN INFRASTRUCTURE PROJECT

Civil engineers are central to the planning. They're traditionally working on the grading and moving of soil. So if our bridge site extends beyond the edge of the shoreline—meaning it's now in water, but we want it on land—the civil engineer will determine how much soil must be taken from one area to shore up the landing.

Civil engineers will be responsible for the actual siting of the bridge, meaning calculating all data to come up with the best physical location. They are finding real-time answers to questions like these: Do we want a very long span, or do we want a short span? What's the best bridge type based on how long it is? Will it have a top and bottom deck? Are trucks going to go over this bridge? Do boats have to go underneath it? Is it navigable by water? How is this going to impact existing roads? What type of vibrations might impact neighboring buildings? All of these considerations will be taken into account to determine the best location for this bridge.

Structural engineers are focused on the actual bridge structure. They are concerned with the steel, the connection points, the concrete needed, how it's reinforced, etc. When the architect presents their design, the structural engineer will figure out the actual weight of the concrete, columns, slabs, and building materials, referred to as the dead load. They're also calculating the live load, snow load, wind load, flood load, seismic load, etc. Structural engineers will apply all these calculations to design a system that will handle the realistic traffic and road conditions to be expected on the bridge without failure or complete collapse.

Mechanical engineers will be responsible for the analysis and eventual monitoring of elements of the bridge, especially connections that could eventually experience corrosion and fatigue. Mechanical engineers are responsible for several of the moving pieces on a bridge, like a lift bridge or gate. They may also be involved in the design for bridge-drainage structures and piping if project details require it.

Electrical engineers and *lighting engineering specialists* will step in to assess lighting options for the bridge. They are considering the type of lighting needed, brightness, cost, etc. This may sound relatively straightforward, but it's not just about determining what's best for the bridge. They have to be cognizant of how that bridge lighting will affect adjacent communities. If an apartment building sits next to the structure, for instance, the residents must be considered before finalizing lighting that may disturb their quality of life.

When the structure is fully designed and the plan is developed, *construction engineers* will price the job with the various contractors so that the agency who will own and operate the bridge is aware of how much it's going to cost.

A *cultural resources engineer* might then be brought in to advise how to handle a stream that will be impacted by this build. What type of fish and small-animal habitats could be disrupted by the foundation that is going in the ground? Are there wetlands that are present that restrict building? The cultural resources engineer identifies and remedies these types of issues so that, finally, the placement of the bridge can be determined.

Once the bridge location is confirmed, regulatory agencies must grant permission for the bridge to be installed at this site. Engineers from within those agencies will now review the plans from a design-and-budget standpoint.

Contractors and on-site engineers, who are responsible for actually building the different aspects of the bridge, are in the field ensuring that its construction meets all quality standards: Are the foundations deep enough at the right locations? Are the spans in fact appropriately laid out? Are connections within the structure proper and to code? There are laboratory tests that are happening for the concrete, steel, timber, masonry, etc. The construction engineer is ensuring that all these materials are safe as the build progresses.

So if one person claims to have designed a bridge, and they're the reason it stands there today, that's untrue. There are so many different aspects of what has to happen that several engineering companies may become involved. Within those engineering companies, there are multiple engineers who are working on the actual project.

EDUCATION REQUIREMENTS
FOR ENGINEERS

The engineering curriculum is math and science rich because engineers use math and science principles every day. It's more like a language we speak, not necessarily something we consciously think about. Some of it is basic math, like the volume of concrete going into a specific location in order to assess the number of trucks needed on-site, and sometimes it's more sophisticated analysis, such as figuring out the probability of rock mass sliding out from an excavation site. Nearly every task associated with a project, from initial stages to final touches, relies on accurate math calculations.

By way of science, we regularly employ the scientific method: we have an idea of what might happen; we gather information; we conduct testing to support our idea; and we come to a conclusion.[48] Science interacts with, supports, and guides engineers every single day.

Chemistry, for example, is regularly used, especially by environmental engineers. They perform soil, groundwater, and air tests to determine if contaminants are present that could affect users of a space. You might be surprised to know that these engineers must also be experts in biology. Why? They apply biosolutions for cleaning soil, such as bugs and bacteria that will consume unwanted contaminants.

Simply put, engineers use math and science to solve problems. Even if you've got the grades, if problem-solving does not appeal to you, engineering is probably not your best career choice.

AREAS OF DISCIPLINE

A complete list of every possible area student engineers can branch into is too large for this book. To get you started, we've prepared a small sample of disciplines under the four main engineering categories, according to Southern New Hampshire University: chemical, civil, electrical, and mechanical engineering.

Types of Engineering Disciplines

As noted on the SNHU website, "There are dozens and dozens of different types of engineering, but when it comes down to the basics, engineering is about using specialized bases of knowledge to solve a problem. Since we encounter a wide variety of problems, we have

---

48   Gavin Wright, "What Is the Scientific Method?," Tech Target-WhatIs.com, accessed September 12, 2023, https://www.techtarget.com/whatis/definition/scientific-method.

an equally wide range of engineering disciplines, many of which are highly specialized and designed to solve those problems.

"In broad terms, engineering can be divided into four main categories—chemical, civil, electrical & mechanical engineering. Each of these types requires different skills & engineering education."[49]

CHEMICAL ENGINEER	CIVIL ENGINEER	ELECTRICAL ENGINEER	MECHANICAL ENGINEER
Process Design Engineer	Structural Engineering	Computer Engineering	Acoustics
Environmental Engineer	Environmental Engineering	Electronics	Aerospace
Plant Process Engineer	Geotechnical Engineering	Instrumentation	Automation
Process Safety Engineer	Transportation Engineering	Optics	Automotive
Technical Sales Person	Water Resource Engineering	Photonics	Autonomous Systems
Environmental Waste Mgmt.	Surveying	Photovoltaics	Biotechnology
Chemical Plant Technical Dir.	Construction Engineering	Power Engineering	Composites
Petroleum Engineer[50]	Municipal Engineering[51]	Radio-Frequency Engineering[52]	(CAD)[53] Computer Aided Design

49   Joe Cote, "Types of Engineering: Salary Potential, Outlook and Using Your Degree," Southern New Hampshire University, August 10, 2022, https://www.snhu.edu/about-us/newsroom/stem/types-of-engineering.

50   UC Riverside Career Center, "Chemical Engineering, Representative Job Titles and Area of Specialization," accessed September 12, 2023, https://careers.ucr.edu/resources/career-planning/careers-in-your-major/chemical-engineering.

51   Fenstermaker Team, "Types of Civil Engineering," Fenstermaker, January 15, 2023, https://blog.fenstermaker.com/types-of-civil-engineering/.

52   TWI Global, "What Is Electrical Engineering? (Definition, Types and Job Salary)," accessed September 12, 2023, https://www.twi-global.com/technical-knowledge/faqs/what-is-electrical-engineering#:~:text=Electrical%20engineering%20is%20now%20split,%2C%20systems%20engineering%2C%20and%20telecommunications.

53   Michigan-Tech, Mechanical Engineers-Engineering Mechanics, "What Is Mechanical Engineering?," accessed September 12, 2023, https://www.mtu.edu/mechanical/engineering.

So any specific area of interest, like the medical, sports, or mechanics industries, can be directed toward engineering because there are likely areas of engineering to support that interest. We encourage students to do their own research to find exactly where they fit in the overall engineering field.

FOCUSING YOUR MAJOR

Typically, in the second semester of senior year, engineering students will participate in a capstone design class. Their university will partner with professional engineers who present information from an actual project to student engineers. The class exhibits their education, knowledge, and skill by designing the project just as if they were on the engineering team ultimately responsible for the build.

Most schools allow students to get input from past engineering professors to get refreshed on concepts they learned in the prior year; however, students are not receiving instruction from their professor during this design. They must work together as a team would in an actual company environment.

> "A capstone course allows college students to demonstrate expertise in their major or area of study. This course is typically required for graduation. Details can vary depending on the major, program and school. Capstone courses typically last at least a semester and sometimes include internships or volunteering. A capstone course typically involves a project such as a final paper, a portfolio, a performance, an investigation, a film, or a multimedia presentation. Some programs use the term 'capstone project' instead of capstone course."[54]

---

54   Ryah Cooley Cole, "What Is a Capstone Course? Everything You Need to Know," Forbes Advisor, updated January 11, 2023, https://www.forbes.com/advisor/education/what-is-capstone-course.

I've worked with the capstone design classes at Howard University and Syracuse University. During our most recent project at SU, the students designed multistory residential and mixed-use structures. The class had the same information the licensed engineers had, minus anything confidential. Students had to work through all that was needed to design the structures, just as I and my team were doing in our office. It was interesting to compare what they came up with to what we actually designed and what was ultimately constructed in New York City.

By the time of the capstone design class, students must have an idea of what engineering focus they ultimately want to work in. For example, I was deciding between structural engineering and geotechnical engineering when I was a senior. When it came to the capstone design class, it was clear that the geotechnical side of the project was where I fit, and I've been working in this discipline ever since.

A NOTE TO PRECOLLEGE STUDENTS REGARDING EDUCATION

Bright students typically get bombarded with questions like this in their eleventh- and twelfth-grade years: "What are you going to study? What do you want to be? Where are you going to go to school?" I got these questions all the time, and I didn't really know what I wanted to do. I was one of the few in my family to pursue college and a professional career and the first to pursue engineering, so my path forward felt a bit unclear.

During high school, however, I took a technical-drawing-type class, and the teacher emphasized the need to apply all that math and science that "we thought we'd never use." Also, the creative aspect of the design had me really enjoying the class. Because of this, when it came time for selecting careers, I actually applied for architectural programs.

Two schools accepted my application for architecture, but Syracuse University, which I wanted to attend, went rogue and offered me a spot in their civil engineering program. I had absolutely no idea what a civil engineer was, but I tried it for a semester with the intention of switching over to architecture the following year. In the very first semester, it became clear that engineers use math and science to solve problems and create things. This fascinated me, so I stayed with it.

At one point—I won't lie about it—I was struggling. The engineering curriculum includes a lot of very difficult classes, even for academically strong students. In the middle of this unstable time, I recall a professor telling me, basically, to just quit. In his opinion, I wasn't going to make it through the courses needed to graduate with an engineering degree. This experience is one of the reasons I speak to and encourage middle schoolers. It's amazing what we can do before someone tells us we can't.

Luckily, during this time, I had real mentors through my family, my friend group, other professors, my fraternity, and my church, and they helped me see that if I really wanted to be an engineer, I would have to intentionally set myself up for success. This was no longer high school, where bright students could get As without really trying. I had to apply myself in a more focused way and study until I fully understood the subject matter.

I eventually became a licensed engineer in fourteen states, and I've received awards for my contributions to the industry. If you are about to walk in this direction, understand that a lot of hard work will have to go in before you get to the fun, like getting muddy on construction sites or seeing massive rigs mobilized to install projects your team has designed.

If any part of facing that much hard work diminishes your willingness to go after an engineering career, it may be best to reassess your direction. If you've got the grades to consider engineering, you likely can go anywhere, so I'd encourage you to find the career you feel passionate about.

If the hard work only makes you more determined to become an engineer—maybe the challenge even excites you—then go after that engineering degree, and never allow anyone to deter you from your path.

MENTORING OR REQUIRED SUPERVISED HOURS

Part of the engineer's license exam includes a design that the student has developed/created and that a supervisor has signed off on. For this to happen, it stands to reason that the new engineer must be working under the direction of a supervising, licensed professional engineer for some time after completing school.

Our industry doesn't specifically refer to supervised hours as mentoring, and to me, there is a strong distinction. The supervisor is required to check the trainee's work and recommend revisions. They aren't necessarily "mentoring." They are there to ensure that the design is correct.

Mentoring is when someone with knowledge, insight, and/or expertise takes time to invest in a less-experienced individual. Is this type of mentoring necessary for licensure? No, it's not. However, from my own observations, an individual can go to the best school, take all the top courses, and make it through in half the time, then step into the field and fizzle out. To successfully become a leader in the engineering space, or any other, education is best complemented by the example of several outstanding mentors.

CHARACTERISTICS THAT FIT WELL IN ENGINEERING

Children bent toward the engineering space may be inclined to intentionally break their toys, try to figure out how they work, and then put them back together again. Drawing is typically a favorite pastime, and their best classes, hands down, are math and science.

The personality geared toward engineering is often very focused, and many engineers perform outstanding work within their team and consciously remain behind the scenes. If that fits your personality and goals, you will likely find a great fit somewhere in the engineering industry.

For those whose personality and goals are geared toward wanting the next level—to earn those promotions and get those high-profile jobs—leadership ability is necessary. In fact, the highly successful engineer typically exhibits these four characteristics or traits in support of their outstanding leadership skills:

1. They Continually Learn and Develop

Continuing education is a requirement for a licensed engineer. Some states are very specific about the number of hours engineering professionals must focus on specific subject matters, such as ethics and/or sustainability.

On a very practical note, I've been an engineer for twenty-plus years, and I've never seen the same project pass over my desk twice. The field requires constant application of what we know and constant learning and development to excel in the next project.

2. They Network

Not every engineer actively networks. As I've already mentioned, many are content to stay behind the scenes. Networking is often a challenge to the engineer personality type because a good number of us are introverts. Leaders, however, very rarely have the luxury of burying their faces in their work. They must deal with others and network to build support around them as they handle their responsibilities.

The unspoken motive in typical networking is that this group is developed to move the networker forward, either in current business or future opportunities. I see it a bit differently.

Networking, to me, has always meant having meaningful friendships within my own peer group. I have close friends from college that I've grown up with. We started jobs, went to one another's weddings, and are raising children all at relatively the same times. Professional connections were never expected of these relationships, but sure enough, a network was created. If I've got an issue, whether personal or professional, I call them; when they've got problems, they call me.

So I'd say networking is very important in general, but my advice is to set out to create quality friendships rather than cold connections.

3. They're Skilled in Organization and Time Management

The reality of an engineer's job is that multiple deadlines are due every week, month, and quarter. If time is not managed effectively, those deadlines can be upon them before they know it, causing a lot of stress and delay for the entire project.

Understanding what aspects of a project are high priority is usually one of the biggest challenges, especially for engineers who are starting out. Leaders in the field develop an organized plan to achieve milestones, and they manage their time so that, as much as possible, the job comes in on schedule and under budget.

4. They Are Confident Public Speakers

Engineers attend a lot of meetings, and we're usually working with a peer group, some subsidiaries, and a project team leader who we report to. In this team environment, it's very important to be able to articulate oneself both in written and verbal form.

Especially when it comes to infrastructure, there will be a number of community meetings to attend before any work begins. Engineers typically are required to speak on a specific aspect of a project in order for financing to be approved. We publish papers and present them at conferences, and we also present internally to our own respective peers. Additionally, when designs are submitted, there are regulatory meetings to attend. These meetings often feature one or more engineers from the team explaining the design to that group.

So public speaking is part of the job. It's not always in front of a podium, in formal wear; sometimes it's just day-to-day reports, but it falls to us quite often to prepare a presentation and deliver it. One of the keys to successfully accomplishing that is understanding your audience. The skilled engineer is able to successfully convey highly technical material to audiences who may not have an engineering background.

THE FUTURE OF ENGINEERING IN HIGHWAY INFRASTRUCTURE

We have existing infrastructure that's aging and failing, and we have a historic opportunity to rehabilitate and recreate it. When we think about the companies that are going to be doing this work, four specific aspects will need to be taken very seriously:

1. Priorities of Upcoming Engineers

The generations of engineers that will be graduating will demand sustainability and equality policies be honored before they'll design that highway, bridge, or park. What type of materials are we using? Where are materials coming from? Are they close by? Are they renewable? Are the specified materials produced by forced labor or modern slavery? Sustainability, plus responsible design and construction, will be something engineers are focused in on.

2. DEI Conscientiousness

In order for engineers to thrive at a company, they'll want to freely bring their full selves to work. Diversity, equity, inclusion—as well as belonging and justice—will need to be prioritized within engineering companies, now and in the future.

3. Hybrid Working Expectations

The upcoming workforce expects hybrid working options. Engineering companies in general will need to figure out how to administrate that in order to hold on to the best and brightest talent. Hybrid work should be balanced with providing meaningful ways for new employees to experience and contribute to the company culture.

4. Artificial intelligence

AI will need to be managed and applied—or rejected—in the field of design. Each engineer with this tool at their disposal must answer questions such as these: Does AI really help us, or does it harm creativity? How can we embrace it strategically? Where should we use it and not use it?

We must tread cautiously with such a tool until we really know what AI means for the design, construction, and engineering worlds.

• • •

As far as the future regarding the infrastructure rebuild, this really isn't as brand new as many folks think. People have been building roads since the days of antiquity; whole people groups and civilizations have had to move from one place to another in a reliable manner. We're going to continue to do that, but the aim will be to do it safer, more economically, and in a way that brings everybody to the table to be a part of those solutions.

7

ARCHITECTS

Career Information

PROVIDED BY PROFESSIONAL ARCHITECT PASCALE SABLAN, FAIA, NOMA, LEED AP

Architect. Activist. Visionary. Leader. Audacious disrupter of the status quo. As an associate principal at Adjaye Associates with over fifteen years of experience, she is the 315th living African American woman registered architect in the United States. Pascale is dedicated to advancing architecture for the betterment of society by bringing visibility and voice to the issues concerning women and BIPOC designers. She founded the Beyond the Built Environment organization to address the inequitable disparities in architecture and serves as the global president of the National Organization of Minority Architects. Pascale has been recognized with several prestigious awards, including the 2021 AIA Whitney M. Young Jr. Award and the Architectural League 2021 Emerging Voices Award.

Pascale has given lectures at colleges, universities, and cultural institutions globally, inspiring future generations to create a more equitable, inclusive built environment. She has been quoted in the *New York Times* regarding her efforts, and *Forbes* magazine described her as "the powerhouse woman," actively changing history with a simple mission: women and designers of color must claim and be credited for their contributions to the built environment.

Her advocacy work in architecture was recently recognized in a feature article on NPR entitled "Very Few Architects Are Black: This Woman Is Pushing to Change That."[55]

About the Career

Definition of Architect (Bureau of Labor Statistics): "Architects plan and design houses, factories, office buildings, and other structures. Architects spend much of their time in offices, where they develop plans, meet with clients, and consult with engineers and other architects. They also visit construction sites to prepare initial drawings and review the progress of projects to ensure that clients' objectives are met."[56]

National Average Salary (Indeed.com): $80,180 annually; Range: $38.55 per hour.[57]

Current State of Architecture in Infrastructure: Needing immediate candidates.

Opportunities to Work for a Company? After college, graduates usually get a job with an architecture company who helps them through the rest of their certification process.

Opportunities for Self-Employment? Yes, but usually architects must be licensed to be in business for themselves. Drawings must be signed and stamped, and this requires licensure. It may not be completely impossible, but it's quite difficult to graduate and go straight into opening your own firm. The exception would be a graduate who is partnering with somebody who's already licensed to help facilitate processes. Most commonly, architects work for a few years to complete their AXP examination process, get a sense of what it is to run a business, and then think about branching off on their own.

55   "Very Few Architects Are Black: This Woman Is Pushing to Change That," NPR, March 12, 2023, https://www. npr.org/2023/03/12/1160836191/black-african-american-architects-architecture.

56   US Bureau of Labor Statistics, Occupational Outlook Handbook, s.v. "Architects," accessed September 12, 2023, https://www.bls.gov/ooh/architecture-and-engineering/architects.htm.

57   BLS, "Architects."

THE CAREER OF ARCHITECTURE

Architects approach most projects with a collaborative mindset, where we're valuing the contributions of consultants, design engineers, and other critical thinkers. We are stewards of the aspirations and concerns of our clients and stakeholders, and we try to manifest our vision for what a great society would need and how our project could support that. That is true for residential, cultural, and other institutional projects, as well as infrastructure projects.

As an illustration, I had the opportunity to work on the Fifth Crossing Bridge in Dubai. The project's trajectory was initiated by collaboration, as the design predominantly focused on comprehending the site and the interplay between the buildings neighboring each bridge landing.

A significant aspect of our creative process involved actively listening to the influential and meaningful aspects of the culture we were designing for. The falcon holds significant importance in the Dubai experience, and we embraced this concept by incorporating it into the tension cables and design elements of the bridge. The curvatures of the bridge landings onto the riverbanks were inspired by the form of a falcon's beak. This concept seamlessly extended to the landscaping design strategy as well, ensuring a cohesive visual theme.

We had to be conscious of specific territorial considerations, given that the bridge landing was situated between the UK Embassy and the US Embassy. Furthermore, since a tram ran along the center of our bridge, it was crucial to coordinate with the public transit municipality to meet their requirements regarding track widths, station size, available resources, and more. This represents the more technical side of an architect's responsibilities, which must be harmoniously balanced with their creative instincts.

Sustainability as well as design excellence were our priorities. On this particular site, wildlife inhabited the waterway. It was necessary to ensure our bridge would not create a negative impact on that wildlife. The entire architectural team had to be vigilant about thinking ahead to how this project would impact the community and surroundings.

Among the more interesting aspects of the details we had to take into account was walkability. Given the extreme heat in Dubai and the regular walking routines of its residents, we had to address the challenge of uphill climbs, especially for the elderly, in such a harsh climate. To tackle this issue, we incorporated air-conditioned rest areas along the bridge, providing spaces where people could take a break, recover, and gather their energy before continuing their journey.

Although the design process for the project was ultimately completed, the project was tabled, and the bridge currently remains unconstructed—unfortunately, this is a disappointment that architects must face sometimes. Nevertheless, take note of the number of pieces the architectural team must keep in order throughout a project. Architecture is not solely about erecting a structure; it entails doing so with the aim of fulfilling the community's needs while upholding sustainability as a core principle.

EDUCATION AND REQUIREMENTS

The majority of pathways toward becoming an architect typically commence with a bachelor's degree in science or architecture, followed by the pursuit of a master's in architecture, resulting in a total of seven years of education. Consequently, many architects possess two degrees. In my case, driven by a long-standing ambition to become an architect, I opted for a streamlined approach. I completed a five-year master's program at Pratt Institute and subsequently pursued my master of science and advanced architectural design at Columbia University, culminating in a total of six years of dedicated study.

Ilana Kowarski, with *US World News and World Report*, explains architectural education requirements in her article "The Two Types of Accredited Architecture Programs":

> "People who want to receive an architecture degree that will facilitate their licensure as an architect have two options: either a five-year undergraduate degree in architecture known as a B.Arch. or a postcollege master's program known as an M.Arch, which usually takes two or three years to complete.

> "Though there are four-year bachelor of science, bachelor of arts, and bachelor of fine arts programs in architecture, they usually are not nationally accredited. Someone with one of these undergraduate degrees may need to attend a master's program in architecture in order to qualify for state licensure.

> "So, before attending an architecture program, any aspiring student should investigate whether that program is accredited by the National Architectural Accrediting Board, or NAAB. A list of all the accredited architecture programs in the US is available on the NAAB website.

> "College hopefuls who are determined to become architects should know that the fastest and least expensive path into the architecture profession is via a five-year undergraduate program, architectural industry experts say. However, those who are unsure about whether they want to pursue this career may prefer to pursue a four-year college degree in architecture or another field, knowing that they can later

supplement their undergraduate education with a master's in architecture no matter their college major.

"Doctoral programs in architecture are available, though that extremely advanced level of architectural education is not at all mandatory to be an architect. It is advantageous, however, for a future academic."[58]

I owe a tremendous debt of gratitude to my mother for her wisdom and foresight. Recognizing the prevailing dynamics within the architectural field, which is predominantly white and male, she strongly encouraged me to pursue the additional credentialing of a master of science and advanced architectural design alongside my degree from Pratt. Her belief was that these dual degrees would provide me with a fairer opportunity to be considered for leadership positions within the industry. I am truly grateful for her guidance and support, as it has undoubtedly shaped my journey as an architect. It is disheartening that such considerations are still necessary in certain fields, but I remain committed to breaking barriers and contributing to a more inclusive and diverse architectural landscape.

PRECOLLEGIATE OPTIONS

Young people who are inclined toward the architecture field are likely artistic, top students, and creative. This certainly was my experience; I wanted to create art in any way that I could.

At about the age of twelve, I was commissioned to paint a mural at the Pomonok Community Center in Queens. As I was painting

---

58   Ilana Kowarski, "How to Study Architecture and Become an Architect," *US News and World Report*, October 2, 2020, https://www.usnews.com/education/best-colleges/articles/what-an-architecture-degree-is-and-how-to-become-an-architect.

this jungle gym with a multicultural theme, someone walked by and said, "Wow, you can draw straight lines without a ruler. That's a cool skill for an architect to have," and they kept walking.

That was the first time that architecture was offered as a potential career pathway. I was always very good in terms of grades in school, and people would always ask me, "Doctor or lawyer, which one are you going to pursue?" Neither, thankfully. Architecture became my answer; I could leverage my artistic talents, capabilities, and love of art into a meaningful career that would enrich the world.

My list of colleges was created solely based on that trajectory, so my mom enrolled me in "What's an Architect?"—a seminar being held at One Penn Plaza, in Midtown. The seminar leaders took us to architecture firms, construction sites, newly constructed projects, and model shops so that we correctly understood what it was to be an architect. It was just a few weeks long, held after school, but I found it to be life confirming in my case. The course provided a non-television-and-movie view of what architects were, and I knew I was destined for this profession.

Students who are thinking about architecture as a possible career can gain invaluable real-world experience by participating in these types of seminars. If I were selecting a course of this type again, I'd likely go through ACE Mentor Program—Architecture, Construction, and Engineering (*acementor.org*) because I'm most familiar with their outstanding curriculum. Similar seminars, however, exist throughout the country. For interested students, Lumiere Education publishes the following list of ten comparable programs:

10 ARCHITECTURE PROGRAMS FOR HIGH SCHOOL STUDENTS

by Stephen Turban

1. Cornell University's Introduction to Architecture Summer Program
2. Hip Hop Architecture Camp
3. Girls Garage's Programs for Girls
4. NOMA Project Pipeline
5. Build San Francisco Summer Design Institute
6. UC Berkeley's embARC Summer Design Academy
7. Columbia University's Summer Immersion: New York City
8. Rhode Island School of Design's Pre-College Summer Residential Program
9. Texas A&M University's Camp ARCH
10. Fallingwater Institute's High School Residencies[59]

MENTORING REQUIREMENTS

Architects must be approved through their respective state and through official mentorship. To become a certified architect, apprentice architects must have licensed supervisors oversee their work and verify that the correct number of training hours have been invested into the required categories. The goal is to ensure that trainees become fully prepared, well-rounded architects. This real-world practice expe-

---

59 Stephen Turban, "10 Architecture Programs for High School Students," Lumiere Education, accessed September 12, 2023, https://www.lumiere-education.com/post/10-architecture-programs-for-high-school-students.

rience supplements the architect's education and prepares trainees for the *Architect Registration Examination (ARE)* process.

"**The Architect Registration Examination® (ARE®)** is a multi-division exam used to assess your knowledge and skills regarding the practice of architecture. The current version of the exam, ARE 5.0, is developed by NCARB and features six divisions. Completing ARE 5.0 by passing all six divisions is required by all US jurisdictions as a key step on the path to earning a license.

"The ARE is designed to assess aspects of architectural practice related to health, safety, and welfare. Specifically, the ARE focuses on areas that affect the integrity, soundness, and health impact of a building, as well as an architect's responsibilities within firms, such as managing projects and coordinating the work of other professionals."[60]

The Architectural Experience Program

"To earn a license and become an architect, you'll need to document real-world experience through the Architectural Experience Program (AXP). Developed by NCARB and required by most US licensing boards, the AXP provides a framework to guide you through earning and reporting your professional experience.

"As you progress through the AXP, you'll build up competency in the skills and tasks you need to practice architecture. With broad experience areas that reflect the current phases of practice, the program prepares you for everything from site design to project management."[61]

The architecture field in general is geared toward continual mentorship. A deliberate goal of industry leaders is to help steward the next generation. To ensure the safety and integrity of building practices,

---

60 National Council of Architectural Registration Boards, NCARB, "Pass the ARE," accessed September 12, 2023, https://www.ncarb.org/pass-the-are.

61 National Council of Architectural Registration Boards, "Gain AXP Experience," accessed September 12, 2023, https://www.ncarb.org/gain-axp-experience.

seasoned architects are required to earn annual continuing education credits. This commitment to ongoing education enables them to stay abreast of the latest issues, materials, and practices that contribute to the creation of safe and sustainable structures.

NATIONAL ORGANIZATION OF MINORITY ARCHITECTS (NOMA)

The National Organization of Minority Architects (NOMA) plays a crucial role in the architecture field by offering mentorship programs and networking opportunities for firms dedicated to supporting the advancement of minority students.

NOMA was founded by many microfirms, meaning staff size was between two and six people. These firms developed out of necessity because, upon graduation, minority students simply weren't hired, regardless of their qualifications. Since they weren't able to gain employment at majority firms, the solution was for these individuals to forge their own paths and start their own practices. NOMA formed to help facilitate the development of these firms.

Throughout the early-to-mid-2000s, the architecture industry, with the help of organizations like NOMA, began to push for more diversity of staff. This resulted in a slow-but-steady rethinking of the way minorities were received into the field. Now, we see the reverse effect: there is a significant reduction of minority-owned firms because many graduates with minority backgrounds are going straight into majority firms.

NOMA support ranges from offering business classes to honing proper procedure to how to cultivate relationships and win substantially sized projects. NOMA offers a platform where small businesses can network, find talent, and pull resources together in order to meet the necessary requirements to qualify for those projects.

NOMA, with a history spanning over fifty years, has been at the forefront of addressing the challenges and burdens of oppression within the architectural profession. As the president of such an esteemed organization, I feel it is a true honor to contribute to its mission and work toward fostering inclusivity, diversity, and equal opportunities within the field of architecture. NOMA's efforts are instrumental in creating a more equitable and representative landscape for architects from underrepresented backgrounds.

CHARACTERISTICS THAT FIT WELL IN ARCHITECTURE WORK

Great architects possess a range of characteristics that enable them to excel in their profession. First, they exhibit strong leadership skills and the ability to coach and guide a diverse team of individuals throughout the duration of a project. They understand the importance of sustaining various aspects of a job and overseeing its different facets to ensure successful completion. Additionally, three specific characteristics strike me as necessary for success in the architectural field:

1. Responsibility
Architects bear a significant responsibility for the health, safety, and well-being of society. They meticulously consider and address every aspect of a design, ensuring that it meets the highest standards of safety, functionality, and aesthetics. Creative problem-solving skills are an essential attribute, as architects are frequently confronted with complex challenges that require innovative and practical solutions.

2. Good Communication

Architects are constantly documenting and explaining their project and process to diverse audiences. They perform presentations and advocate to consultants, clients, communities, agencies, building departments, etc. Effectively communicating their message and securing project funding in the field of architecture necessitates strong public speaking skills and the ability to adapt your message to different platforms. Architects aim to be proficient in conveying their ideas, vision, and design concepts to a diverse range of stakeholders, including clients, investors, and the public.

A high level of comfort with technology is required as well, as different software tools are constantly implemented as teams articulate ideas and collaborations progress.

3. Fiscal Awareness

Architecture, being a creative field, often poses challenges when it comes to persuading clients about the value of the work being done. Consequently, architects often adopt a fiscally responsible approach, recognizing the significance of the effort, energy, time, and resources invested in a project. They understand the need to balance artistic vision with practical considerations, ensuring that the design not only fulfills the client's requirements but also aligns with their budgetary constraints and expectations.

THE FUTURE OF THE ARCHITECTURE FIELD

A significant point to consider is that a lack of infrastructure has ramifications and will impact how communities are able to thrive. Case in point, there are some communities that still don't have access to mass transit and Wi-Fi. This puts them at a disadvantage, especially when it comes to job applications and obtaining other critical online services.

Whether infrastructure causes smog to hover over minority neighborhoods or waste treatment facilities are concentrated in communities of color, infrastructure has been used to facilitate racism and oppression in the built environment. Thoroughfares and avenues were intentionally erected to cut down communities and segregate them from beautiful spaces and resources throughout a city. So when we think about repairing or redesigning our infrastructure, we must be strategic in creating infrastructure that manifests aspirations and design justice.

The restructuring of our highway system will require architects who understand the original damage the highway system created. Financial expense should not be the determining factor on whether some of these structures should be corrected by a complete tear-down and rebuild. We've invested millions of dollars—if not billions—in creating architecture that oppresses; why not leverage an equal amount of money to correct it?

The best path forward is to speak with the respective communities who have sustained this damage and understand where the challenges are. What spaces do they need access to? How can new roads and/or highways be designed to facilitate that? Some examples of solutions are projects where the space underneath highways and overpasses is repurposed to allow for cultural spaces and community programs to occur. Landscaping can be manipulated to create more of a threshold into community services rather than a barrier.

The future of the profession is one that eradicates racism, sexism in all forms, and oppression. This includes the way architects practice and also the projects we construct and build.

PART 3

CHOOSING YOUR LANE: THE REAL ESTATE SECTOR

RIGHT-OF-WAY, APPRAISAL, AND TITLE FIELDS

8

THE RIGHT-OF-WAY OCCUPATION

About the Career

Definition of ROW Agent (*Dictionary of Occupational Titles*): "Title(s): Right-of-Way Agent (any industry) alternate titles: claims agent, right-of-way; permit agent; negotiates with property owners and public officials to secure purchase or lease of land and right-of-way for utility lines, pipelines, and other construction projects: Determines roads, bridges, and utility systems that must be maintained during construction. Negotiates with landholders for access routes and restoration of roads and surfaces. May examine public records to determine ownership and property rights. May be required to know property law."[62]

Current State of ROW Work in Infrastructure: Needing immediate candidates!

Opportunities to Work for a Company? Yes; that's typically how someone gets work. An agent employs a project manager, and they will supervise the specific job ROW teams are assigned to. We follow our agency's policies and a manual from the commonwealth.

Opportunities for Self-Employment? If ROW agents can meet all the requirements of the commonwealth or agency, yes, there are opportunities for self-employment. Likely, this person will take their schooling to the senior agent level. The requirements and paperwork for self-employment is complex and extensive, so self-employment in the ROW field is best for those with a highly meticulous mindset.

THE RIGHT-OF-WAY CAREER

In most any type of infrastructure build, there will be landowners who find that their private property sits in the direct path of one of the roadway revitalization projects. Maybe a storefront will lose its parking lot to an expanding road; a family's private residence may have to be razed to make way for a new neighborhood street that connects to a major highway; the alleyway behind a busy office park may have to become an access point and staging area, causing interference with the tenants' ability to conduct business. Right-of-Way

---

Dictionary of Occupational Titles, CODE: 191.117-046, accessed September 12, 2023, https://occupationalinfo.org/19/191117046.html.

(ROW or R/W) experts are assigned to intermediate between those building infrastructure and those who own the land needed.

The term "ROW expert" is a general term. Under this category are two specific areas of expertise: agents and professionals. A right-of-way *agent* and a right-of-way *professional* are similar in that they both deal with the acquisition, utilization, and management of rights-of-way cases.

A ROW *agent* is typically responsible for negotiating and acquiring the necessary rights-of-way for a project or development. They work closely with landowners, government agencies, and other stakeholders to secure the required access and easements. They may also handle the documentation and paperwork associated with the process.

A ROW *professional* may not only negotiate and acquire rights-of-way but also manage and maintain them over time. This could involve ensuring compliance with legal and regulatory requirements, coordinating with other professionals such as surveyors and engineers, and mitigating any potential conflicts or disputes that may arise.

ROW *procedures* deal specifically with the eminent domain laws that come into play with infrastructure projects. Specific procedures have been put in place that must be strictly followed in order to ensure each party in this situation is treated as fairly as possible. Under the general theory of eminent domain, the government has the right to take a person's private property in furtherance of a public works project, as long as the owner is fairly compensated.

Eminent Domain

"The power to take private property for public use by the state, municipalities, and private persons or corporations authorized to exercise functions of public character. Housing Authority of Cherokee National of Oklahoma v Langley, Okl., 555 P.2d 1025, 1028. Fifth Amendment,

US Constitution. In the United States, the power of eminent domain is founded in both the federal (Fifth Amend.) and state constitutions. The Constitution limits the power to taking for a public purpose and prohibits the exercise of the power of eminent domain without just compensation to the owners of the property which is taken. The process of exercising the power of eminent domain is commonly referred to as 'condemnation' or 'expropriation.' The right of eminent domain is the right of the state, through its regular organization, to reassert, either temporarily or permanently, its dominion over any portion of the soil of the state on account of public exigency and for the public good."[63]

So let's say PennDOT is widening a main thoroughfare in Pittsburgh. All the designs are complete; it's gone through years of approvals—because ROW experts don't become involved until all engineering and architect design contributions are completed; then a plan is created.

That plan must be reviewed by the environmental engineers to clear any ecological issues before it's finalized and the agency has it in hand. When they receive it, PennDOT clearly sees that acquisitions will be necessary, meaning private landowners own part of the property needed to complete the road widening. The ROW professional is called in, and the process of resolving these conflicts will go something like this:

63   *Black's Law Dictionary*, s.v. "eminent domain," accessed September 12, 2023, https://blacks_law.en-academic.com/9012/eminent_domain.

A plan arrives from an agency engineering consultant; it provides all the acquiring data such as ownership, deed, size of acquisition, and the particulars needed to start the process of claimant notifications.

A drive-by of the property is organized to verify plan data.
(We travel regularly to appointments, sometimes two hours or more away.)

A field-view report is written up. We note any changes to the landscape that we see and make any necessary recommendations.

If a title company is not being retained, we do the abstract (search) work.

Files are created, and we prepare all needed documents.

All documents are meticulously peer reviewed.

Notification of the eminent domain acquisition is sent to the landowner. This means the landowner is alerted that their property is about to be acquired.

When advised to proceed, we await the appraisal before we commence actual negotiations.

We have timelines and deadlines to meet before eminent domain proceedings begin. At the same time, of course, we're busy with other tasks such as photographing possible needs for appraisers or stopping at the county courthouse to record deeds.

Most landowners are happy with their settlements; if this is the case, the contract goes through, and the file is closed for another successful eminent domain acquisition.

However, if we cannot amicably or administratively settle a claim, then it is referred back through the chain-of-command of the agency condemning it.

Paperwork for a condemnation case is sent to the claimant and their attorneys, if they've retained representation.

Parties are advised that eminent domain, or condemnation action, will be taking place.	If landowners will not settle, the property is condemned.

A board receives the claim and schedules an in-field review of the condemnation details with the claimant and the acquiring agency.
At this point, a determination is made, pro or con, and all parties must comply with it.

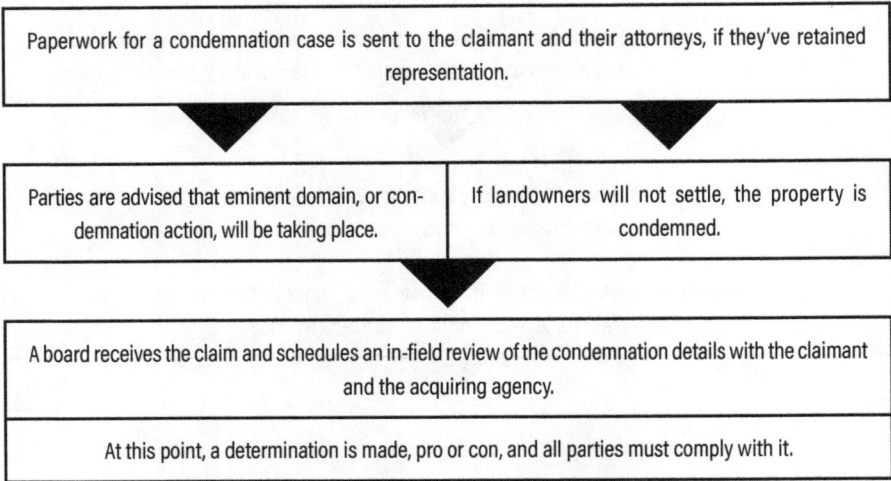

"Eminent domain and condemnation go hand in hand. While eminent domain refers to the ability of the government to take private property for a public use, condemnation is the process by which that happens."[64]

Eminent domain laws can fall under the purview of the federal government or an individual state, and state laws vary dramatically. ROW experts generally understand that the ROW system is not perfect, but eminent domain law in the United States dates back to the 1800s, and it's our job to adhere to it.

The US Supreme Court first examined federal eminent domain power in 1876 in Kohl v. United States, 91 US 367, 371 (1875). This case presented a landowner's challenge to the power of the United States to condemn land in Cincinnati, Ohio, for use as a custom house and post office building. Justice William Strong called the authority of the federal government to appropriate property for public uses "essential to its independent existence and perpetuity."

---

64   LaPonsie, "Basics of Eminent Domain and Condemnation."

The Supreme Court again acknowledged the existence of condemnation authority twenty years later in United States v. Gettysburg Electric Railroad Company, 160 US, 668, 679 (1896). Congress wanted to acquire land to preserve the site of the Gettysburg Battlefield in Pennsylvania. The railroad company that owned some of the property in question contested this action. Ultimately, the court opined that the federal government has the power to condemn property "whenever it is necessary or appropriate to use the land in the execution of any of the powers granted to it by the constitution."[65]

SALES AND REAL ESTATE A PLUS

For those wanting to pursue this career, sales experience, especially real estate sales experience, will give them an edge over any competition who lacks this background. At one point, the company I work for came under new management. When they tried to increase the number of ROW professionals on staff, they hired bright individuals, but none of them had a sales or real estate background. I was in the field with them every day, and they just lacked a certain perception of what they were doing that was hard for them to catch on to. They had training materials at their disposal; all processes were reviewed before approaching a landowner; role play was even rehearsed. Yet they still exhibited a lack of confidence in what they were explaining during acquisitions.

When a representative can't answer a mildly anxious claimant's questions, it reflects poorly on the entire acquisition. Worse, it can cause a lack of trust in who we are at a time when we're striving to build confidence with the landowner.

---

65 US Department of Justice, "History of the Federal Use of Eminent Domain: Early Evolution of Eminent Domain Cases," accessed September 12, 2023, https://www.justice.gov/enrd/history-federal-use-eminent-domain#:~:text=The%20U.S.%20Supreme%20Court%20first,house%20and%20post%20office%20building.

When these trainees did not work out, a new round of people were hired who were required to have some sort of real estate background, whether in sales, title work, or legal. It was a night-and-day difference. These trainees instantly understood the application of law because they came to the job already familiar with many real estate transfer protocols.

When I took this career on, after years in sales and sales management, I found it rather easy to transition. It's about relating to people while guiding them, and in this situation, they are in serious need of assistance. Landowners often think they understand what's going to happen in the ROW acquisition process the very first time the ROW team speaks with them. Then, a tidal wave of questions forms and may not let up the whole way through. The ROW professional must know the eminent domain laws in their jurisdiction and be able to answer any questions accordingly. Those with a sales background—especially in real estate sales—seem to understand this more readily.

EDUCATIONAL REQUIREMENTS FOR ROW AGENTS

To perform the tasks of a ROW professional, a degree is not necessarily required, in my opinion. I've seen degreed individuals fail at the job and nondegreed individuals succeed. That said, the better educated one is, the easier it will be to grasp the legal concepts. Also, individual employers may require a degree if you don't bring several years of prior experience to the job.

Negotiators generally are not overly paid or very wealthy. If an individual is looking for the higher wages that can come from this profession, they'll need a senior IRWA credential. Senior agents are in demand, and their pay grade typically reflects this. For those seeking the ROW senior agent designation, classes are offered by the International Right of Way Association (IRWA), the National Highway Institute (NHI), and Highway Infrastructure Courses Online.

International Right of Way Association (IRWA)	irwaonline.org
National Highway Institute (NHI)	nhi.fhwa.dot.gov/home.aspx
Highway Infrastructure Courses Online	highwayinfrastructurecourses.com

MENTORING OR REQUIRED SUPERVISED HOURS

There are no mentoring requirements for ROW agents; in my experience, the project managers tend to mentor continually. Someone interested in this field who is looking for guidance should seek the advice of a ROW administrator from their local highway agency such as the DOT or Highway Commission. My best advice is to call this administrator and introduce yourself; ask them some questions; show a sincere interest. Leaders in this industry—like most others—are willing to help somebody who is genuinely interested in the job.

To find these individuals, simply search for them on the internet. In your search bar, type in, "right-of-way administrator in *(your state)*." This simple search usually brings up phone numbers and other relevant contact information.

CHARACTERISTICS/SKILLS THAT FIT IN THE RIGHT-OF-WAY FIELD

The agencies that ROW teams work for build highways, roadways, bridges, etc. Just as the engineers and architects/designers have rules regarding the design, there are rules regarding eminent domain that must be followed on-site. ROW experts are brought in to ensure that those rules are being complied with.

Now, this can be an in-your-face confrontation, but combativeness does not get the best results. To really do the job productively, there are eight character traits/skills that work to create success in the ROW field:

1. Leadership Ability

Engineers see things one way, the homeowner sees it completely differently, and the ROW professional is stating the rules/laws to the situation. We are hired to explain the rules to both parties, and sometimes we must insist they be followed. It's often the case that on-site personnel need reminding of what the law states and how much compensation is allowable for landowners' losses. If we're challenged, we have to stay rock solid and literally show them the code. We're not advocates; we work for an agency, and we have a manual and a policy we must follow.

The ROW professional must hold on to the authority that belongs to them in a scenario where they're often seen as an outsider who is "interrupting the flow of the job." All of this takes a certain measure of leadership ability and backbone to accomplish. If the ROW trainee doesn't understand their purpose on-site and/or they cower during confrontations, the leadership skills needed for the job are lacking.

2. Diplomatic Negotiating Skills

When you're telling someone that an entity is entitled to property they've worked hard for, diplomacy, understanding, and consideration of the homeowner's position is necessary. I have seen some who approach the job like a government employee, just coldly announcing to the homeowner that they're about to lose their property. As you can imagine, landowners generally don't respond well to this style, and they often become instantly combative.

A little diplomacy changes that scenario into a completely different conversation. First, explaining the need for the build and how it will benefit their community paves the way for the news that their property is needed. Requesting their cooperation as the process engages has a more calming effect than trying to throw them into harsh compliance. The most successful approach I've seen is to try to make sure they understand all the facets of their acquisition process, to *avoid* resistance that can hold up progress.

3. Fair-Minded

I can't necessarily speak for anyone else in this industry, but I and the company I work for take full advantage of everything the law provides for the claimant. We advise them, and continually remind them, of all the compensation they can apply for. If something needs to be addressed, the ROW team attempts to answer it or involve an engineer to see if any concerns can be alleviated.

If we cannot accommodate the landowner's requests, it's normally a matter of satisfying them with a reasonable claim settlement. If they still refuse to cooperate, our next course of action is to enforce a declaration of taking and let the respective agency condemn the property.

> **5-15.512: Declaration of Taking Act**
>
> "The Declaration of Taking Act (see 40 U.S.C. § 3114) authorizes the United States to acquire an interest in land immediately upon the filing of a declaration of taking with

a court and the deposit in the court of the estimated compensation stated in the declaration."[66]

We don't want properties to fall into condemnation, but in a case of utter resistance, there is no other choice under the law. We must rely on the rules of our job and a fair-minded approach to move the landowner forward.

4. Attention to Detail

ROW work comes with an immense amount of paperwork and many documents to differentiate from. In all honesty, had I known about this element of the job, I probably wouldn't have taken the career on.

There is paperwork when you receive the plan; when the appraisal is completed; when you send out notifications; and when an offer is made. Depending on the situation, we file the request for the declaration of taking, closing statements, sales agreements, temporary construction agreements, various bank forms, titles, and deeds, and again, depending on the scope and type of acquisition, up to ten additional documents may apply to each individual situation.

Organization and attention to detail, if they aren't natural skills, will need to be developed in order for a ROW professional to succeed.

5. Computer/Technological Aptitude

Different components of the ROW job require familiarity with different computer programs. For instance, we may review plans in AutoCAD, type a report in Word, then jump into various communications platforms to keep information accurately updated between project consultants and respective agencies.

---

66   US Department of Justice, 5-15.512: "Declaration of Taking Act," accessed September 12, 2023, https://www.justice.gov/jm/jm-5-15000-land-acquisition-section#:~:text=The%20Declaration%20of%20Taking%20Act,compensation%20stated%20in%20the%20declaration.

The real challenge for the ROW professional is not necessarily that they have their own complex program to learn for their field; it's that they must work within the parameters of the software chosen by each agency, and they're all different. Familiarity with different technological components becomes critical to stay on pace. A high degree of comfort flowing between computer platforms, and seamlessly adjusting to big and small nuances about them, will be necessary.

6. Math Aptitude

ROW professionals use math every single day, all day, for making informed decisions, evaluating financial aspects, and ensuring compliance with legal requirements. Without mathematical calculations and concepts, we wouldn't function with any reliable accuracy.

When acquiring property through eminent domain, ROW professionals must calculate the amount of compensation to be provided to the property owner. This involves considering factors such as property value, loss of use, and potential damages, which are determined using mathematical calculations. We often work with surveyors to determine property boundaries and legal descriptions. This involves using geometric principles, trigonometry, and other mathematical concepts to accurately measure and map out the land.

Further, ROW professionals may be involved in budgeting and cost analysis for infrastructure projects. This requires mathematical skills to estimate expenses, evaluate financial projections, and assess the overall feasibility of a project.

7. Comfort with Strict Procedure

ROW work requires a military-style level of functioning. We follow procedures and guidelines, reporting to a project manager daily. Only one voice reports to the chiefs as we follow a chain of command.

After being a self-managed salesperson prior to taking this career on, I found that the rigid chain of command was, personally, the most difficult aspect of the job for me.

8. Writing Ability

Included with the mountainous paperwork comes the need to write reports regarding field observations, recommendations, etc. ROW professionals must be aware that any claim can end up in court with their writing being a focal point of the case. Adopting a more exacting and precise way of explaining oneself on paper is really the best style for those pursuing this career.

THE FUTURE OF THE ROW OCCUPATION IN HIGHWAY INFRASTRUCTURE

Locating the statistics on how much US land is owned by the citizenry was particularly difficult. We were finally able to trace a piece written by Gene Wunderlich that goes back to November of 1978, and even then, it seems he had difficulties locating these statistics. This was his determination, however, after much research:

"We can only generally characterize U.S. landownership. The Federal Government owns about 33 percent of the 2.3 billion acres; private individuals own 60 percent; State and public agencies and American Indians own the rest."[67]

So think about that: Roughly 60 percent of the land where infrastructure reform may take place is owned by someone. *Any* of these projects that require private land will send ROW experts spinning with work.

---

67   Gene Wunderlich, "Facts about US Landownership," November 1978, accessed September 12, 2023, https://www.ers.usda.gov/webdocs/publications/41882/30067_landownership.pdf?v=41143.

For instance, there is currently an eminent domain battle going on in Pittsburgh over a $2 billion clean water plan.

"The $2 billion Clean Water Plan is being designed in order to eliminate 7 billion gallons of an estimated 9 billion gallons of untreated wastewater that goes into the Allegheny, Ohio, and Monongahela rivers each year due to stormwater runoff in the Pittsburgh region.

"With such a big project, however, there's a need to buy up many properties along where the tunnels will be located—and that has raised concerns in the community.

"So far, Williams said Alcosan has bought six properties and is negotiating with several others for near surface needs. It's a process she said Alcosan has allocated a year's time to secure the 'needed near surface' properties, and it includes outreach, surveys, appraisals and reaching a settlement, with eminent domain only one of the mechanisms used."[68]

So there is controversy right at the time of this writing, over land needed for clean water. As more and more efforts are put into clean-water solutions, Wi-Fi access, and connecting communities, it seems clear that the future of ROW work is likely to need thousands of experts in the field to accommodate the number of potential projects.

The BIL looks toward a well-needed, rebuilt American highway system; talented negotiators are a necessary part of our infrastructure, and they'll be in serious demand as the BIL's goals progress. There's not only a future for those pursuing a career in this industry but also the very real chance that you'll be effecting a better future for all of us.

---

68   Tim Schooley, "Fighting Eminent Domain," July 27, 2023, Pittsburgh Business Times, accessed September 12, 2023, https://www.bizjournals.com/pittsburgh/news/2023/07/27/fighting-eminent-domain-alcosan-tunnel-project.html.

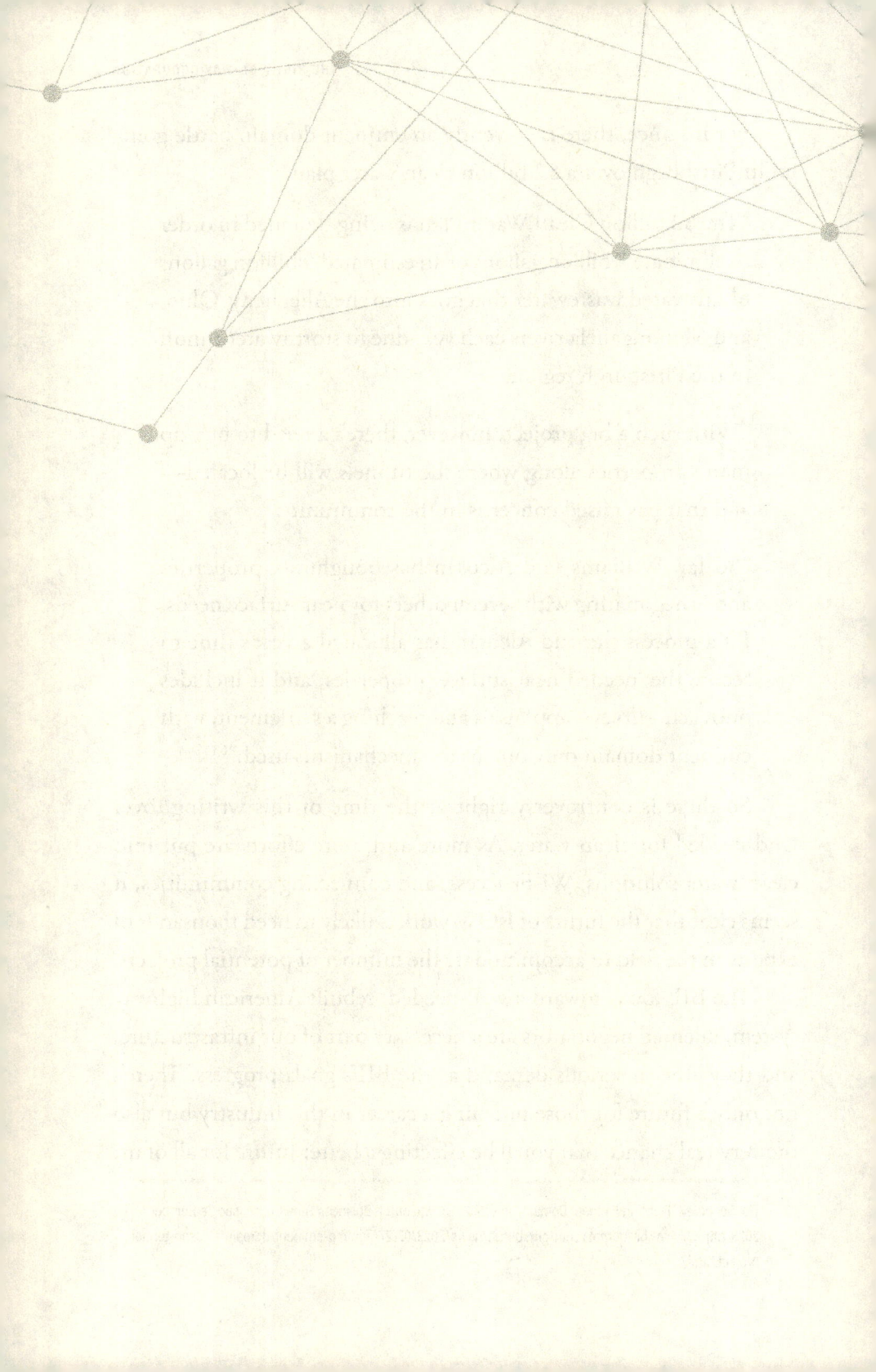

9

APPRAISAL WORK

Career Information

PROVIDED BY PROFESSIONAL APPRAISER DEREK MOLEN, R/W-AC, SRA, CDEI

Derek R. Molen, R/W-AC, SRA, is vice president of Vista Realty Services, Inc. He specializes in appraisal, appraisal review, and valuation services relating to acquisitions and dispositions of property on behalf of clients, including public agencies, quasi-governmental entities, municipal authorities, attorneys, and private property owners, among others. Mr. Molen has worked in the real estate field since 2005, and he is currently credentialed as a certified general appraiser in eight states. His experience involves a variety of assignments, such as condemnation, eminent domain, right-of-way acquisitions, project scoping, excess property dispositions, conservation, and litigation. He can be reached by email at derek@vistarsi.com. As a self-described "real estate geek," he welcomes your inquiries!

About the Career

Definition of Appraiser (Bureau of Labor Statistics): "Property appraisers and assessors provide a value estimate on real estate and on tangible personal and business property."[69]

---

[69]   Bureau of Labor Statistics, "Appraisers and Assessors," accessed September 12, 2023, https://www.bls.gov/ooh/business-and-financial/appraisers-and-assessors-of-real-estate.htm.

National Average Salary (Indeed.com): $57,606 per year. Range: $14,000 to $142,000 annually.[70]
Current State of Appraisals in Infrastructure: Needing immediate candidates.
Opportunities to Work for a Company? There are many companies who hire appraisers, and many will sponsor candidates through their general appraisal training.
Opportunities for Self-Employment? Yes. In fact, many appraisers are either very small shops or one person, with a home office they work from.

THE CAREER OF APPRAISAL WORK

At the point that the appraiser gets involved in highway infrastructure projects, the ROW agent has already assessed the area and sent their report to the agency, and they now need to know the value of the acquisitions so they can negotiate a fair price with the owners. Appraisers assess the fair market value of eminent domain properties to ensure fair compensation is provided to property owners. This includes properties that are acquired for the construction of new highways or those that may be impacted by changes to existing highways.

Appraisers may also be involved in assessing the value of properties adjacent to highways. This can include determining the impact that highways may have on property values and helping stakeholders make informed decisions regarding development, investment, or potential mitigation strategies. For instance, if a multiuse plaza was being built on relatively open land, but a roadway, hospital, and business park are in development, the value of that land may change dramatically. Appraisers step in to determine this change in value.

---

70   Indeed Editorial Team, "Average Real Estate Appraiser Salary at Each License Level," Indeed, accessed September 12, 2023, https://www.indeed.com/career-advice/pay-salary/how-much-do-real-estate-appraisers-make.

For the most part, however, the infrastructure rebuild will specifically require eminent domain appraisals.

> "Eminent Domain Appraisals are most often developed as a result of a whole or partial taking of real estate by a government agency or utility company. The purpose of an Eminent Domain Appraisal is to determine a fair amount of compensation to pay a property owner experiencing a loss of property or property rights."[71]

APPRAISERS VERSUS ASSESSORS

One aspect of the career that confuses people is the difference between appraisers and assessors. I can't even tell you the number of times people have referred to their appraisal as an assessment and vice versa. Scott Steinberg, keynote speaker and best-selling author, offers a succinct explanation to distinguish between the evaluations conducted by appraisers and assessors:

> "Appraised value estimates a property's general worth as determined by a home appraiser and is used in the mortgage approval process. On the other hand, the assessed value is determined by local tax assessors and affects how much you'll pay in property taxes. Both are ways to determine the value of a home."[72]

---

71   Elliott & Co. Appraisers, "Eminent Domain Appraisal," accessed September 12, 2023, https://elliottco.com/services/appraisals/eminent-domain-appraisal/.

72   Scott Steinberg, "Understanding Appraisal vs. Assessment," Quicken Loans, Updated Sept. 13, 2023, https://www.quickenloans.com/learn/assessed-value-vs-appraised-value

"The Bottom Line:

Your home's appraised value effectively reflects what you might expect to get in exchange for the sale of the property if you put it up at market. Its tax-assessed value is instead used to determine how much you can anticipate paying each year in property taxes." [73]

APPRAISERS

When we get into the actual appraisal field, there are two licensing categories that the work involved falls into: residential appraisals and general appraisals.

Residential Certified Appraiser versus General Certified Appraiser

"The 'Certified Residential Appraiser' and 'Certified General Appraiser' may sound similar to many people. It seems that both have a degree of recognition in the real estate market since they hold the 'Certified' status. But if you are thinking of pursuing a career in the appraisal field, you better figure out their distinctions before deciding which license to obtain.

"The Certified General Appraiser has a broader scope of work than a Certified Residential Appraiser.

"This is the major difference between Certified Residential Appraiser and Certified General Appraiser: Although you can appraise any property regardless of its value or complexity, it is restricted to properties with up to 4 dwelling units.

---

73   Victoria Araj, "Understanding Appraisal vs. Assessment," Rocket Mortgage, updated Jun 2, 2024, https://www.rocketmortgage.com/learn/appraisal-vs-assessment.

"In contrast, a Certified General Appraiser can appraise any real estate. In addition to residential units, CG appraisers can evaluate for all different types of properties. For example, commercial buildings, retail stores, hotels, golf courses, industrial plants, farmland, schools, and even cemetery."[74]

Residential property appraisers are critical, but exclusively appraising residential homes for mortgage lenders limits available income opportunities. Notably, a general appraisal certification allows for residential appraisals.

"The next level of real estate appraiser licensing is the certified residential appraiser. On average, they earn approximately $10,000 more per year than licensed appraisers. The highest level of appraisers, certified general appraisers, earn about $15,000 more per year than certified residential appraisers."[75]

In my opinion, if it's at all possible, up-and-coming appraisers would be wise to pursue the general certification. There are so many more opportunities for substantial income with that licensure.

TYPES OF APPRAISAL WORK

So the overall appraisal field consists of residential appraisers and general appraisers. The field is further broken down by the type of work being appraised. When most people hear the word *appraisal*, they automatically think of appraisals for lending purposes, such as buying a house. Some of the most lucrative appraisal jobs, however, are not related to mortgage lending. For example, appraisals are often

---

74   Jacob Coleman, "What Is the Difference between a Certified Residential Appraiser and a Certified General Appraiser?," Real Estate Career HQ, accessed September 12, 2023, https://realestatecareerhq.com/difference-between-certified-residential-appraiser-and-certified-general-appraiser.

75   Indeed Editorial Team, "Average Real Estate Appraiser Salary at Each License Level," updated March 10, 2023, https://www.indeed.com/career-advice/pay-salary/how-much-do-real-estate-appraisers-make.

needed for divorces, estates, tax assessment appeals, eminent domain cases, and for many other nonlending purposes. Below are some examples of niche spaces appraisers can move into.

Types of Appraisal Work		
Arbitration	Depreciation Deductions	Inverse Condemnation
Bankruptcy	Divorce	Litigation Appraisals
Business Appraisals	Eminent Domain	Mediation
Commercial Loans	Environmental Issues	Partition Lawsuits
Construction Defects	Estate and Gift Tax	Real Estate
Consulting	Foreclosures	Residential (Homes)
Contracts	Fraud & Misrepresentation	Tax Assessments

EDUCATION AND REQUIREMENTS FOR THE GENERAL APPRAISER

To become certified at any level, there are classroom training hours and in-field, supervised-training requirements that work together to prepare the general appraiser for licensure. The exception to this is the very first seventy-five hours of coursework (Appraisal Principles, Appraisal Procedures, and fifteen-hour Intro to USPAP). These first seventy-five hours qualify you to *become a trainee*, and they are strictly classroom hours, so there is no supervisory requirement for completion.

Moving on to earning the licensed residential appraiser credential will require an additional seventy-five hours of classroom training and one thousand hours of supervised training. This is where you will need a supervisor/mentor in order to fully complete your training. Earning classroom hours in tandem with in-field supervised hours is the best way to reinforce correct procedures.

The next certification to achieve would be the certified residential appraiser credential. This requires an additional fifty hours of classroom training (for a total of two hundred hours) and fifteen hundred hours of supervised training.

After the CRA credential, trainees are finally on the last leg of their journey to certified general appraiser. The CGA credential requires an additional one hundred hours of classroom training, for a total of three hundred hours, and three thousand hours of supervised training.[76]

THE BACHELOR'S DEGREE REQUIREMENT FOR CGAS

To begin the first seventy-five hours to be a trainee appraiser, all that is required is a high school diploma. However, one of the big hurdles to folks obtaining the certified general appraiser license is that this path requires a bachelor's degree. To pursue the appraisal field specifically for the highway restructuring authorized by the BIL, appraisers need to take their education all the way to certified general appraiser, as that's where the demand will be. Before most states will put an appraiser on their department list for upcoming appraisal jobs, the CGA certification is required.

It is worth mentioning that this can be done while earning experience hours, meaning you can be going to school while working on the three thousand hours of supervised training. Keep in mind the pursuit

---

76   "Appraisal Institute Certified General Real Property Appraiser," Appraisal Institute, accessed September 1, 2023, https://www.appraisalinstitute.org/education/upgrade-your-career/commercial-path.

of this career will require steps and enough time to accomplish them. One of the first is to obtain your bachelor's degree.

SCHOOLS FOR APPRAISAL EDUCATION

The CGA will obtain their bachelor's degree from the accredited college of their choice; however, the appraisal coursework, which actually trains the CGA for what they'll be doing on a daily basis, comes from a specialized appraisal trade school.

The top two schools for receiving your appraisal education are the Appraisal Institute and McKissock Learning. Both of these schools can get you from zero to certified with their programs.

The International Right of Way Association (IRWA) offers additional specialized coursework, although not all classes required for certification are offered by the IRWA.

A fourth national organization that offers the complete appraisal coursework is the American Society of Farm Managers and Rural Appraisers (ASFMRA). Government agencies, such as the Department of the Interior and the Bureau of Indian Affairs, use appraisals for various functions, and a lot of ASFMRA folks specialize in this type of work.

Schools for Appraisal Education	
Appraisal Institute	https://www.appraisalinstitute.org
McKissock Learning	https://www.mckissock.com/appraisal
International Right of Way Association (IRWA)	https://www.irwaonline.org/courses
The American Society of Farm Managers and Rural Appraisers (ASFMRA)	https://www.asfmra.org/education

All coursework must be completed before a candidate can sit for the general certified exam; the exam is four to six hours long, and the first-time pass rate is somewhere in the 55–65 percent[77] range—so it's a pretty daunting exam for many applicants.

CERTIFICATION: KNOW YOUR STATE'S LAWS

The Appraiser Qualifications Board puts out the standards for appraisal certification at a national level. The states must follow those minimums, but they're free to add on to those requirements if they choose. For example, the AQB states that an appraiser must be certified for three years before they can take on a trainee; Pennsylvania has more stringent guidelines, and their requirements state that an appraiser must be certified for five years before taking on any trainees. Each state has different requirements, so it's important to know your state's rules. Below, we've included two trusted resources for finding the requirements for specific states.

The Appraisal Foundation	https://www.appraisalfoundation.org/imis/TAF/ Standards/TAF/Standards_ Qualifications.aspx?hkey=f95f32ad-67dc-439a-b82b-6bf3ea89fa44
The Appraisal Institute	https://www.appraisalinstitute.org/become-appraiser

PRO TIPS REGARDING COURSEWORK

In my estimation, one of the biggest challenges that aspiring appraisers face is actually getting jobs after their coursework is complete. Because

77   National Fair Housing Organization, "Barriers to Entry into the Appraisal Profession," accessed September 12, 2023, https://nationalfairhousing.org/wp-content/uploads/2022/03/PartIII_section5_NFHA-et-al_Analaysis-of-Appraisal-Stndards.pdf.

finding a mentor can be so difficult, tip number one is to take that 75 hours of pretraining license classes but not necessarily roll them into the next 225 hours of coursework until you've secured a mentor. To really grasp the higher-level coursework, it's beneficial to have the hands-on experience that comes with working with your supervisor.

Tip number two is in regard to realistically planning a training-completion date. CGA guidelines call for an eighteen-month minimum on obtaining your three thousand supervised hours.[78] This means certification cannot take place before eighteen months have passed, even if you were able to get that many training hours under your belt.

I want to really caution anyone against thinking that the path to general certification will be a quick eighteen-month course. To obtain enough training hours in that short a time span would require logging forty hours a week for a solid eighteen months. I don't know anyone who's been able to obtain that many logged experience hours in such a short period of time. For most people, realistically, the path to becoming a certified general appraiser is more like a four-to-six-year process.

THE REAL ESTATE CONNECTION

My first career, right out of high school, was as a real estate agent, and I did that for about ten years. I found the sales involved in the job wasn't the best personal fit for me, but the time researching at my desk was, so I stuck with it. As I look back, I'm glad I did because the field of real estate is what opened doors to get me into the appraisal field. As I became more familiar with the appraiser's role, I knew I wanted to move from real estate sales into appraisals.

---

78   Appraisal Institute, "Certified General Appraiser, Step 3," accessed September 12, 2023, https://www.appraisalinstitute.org/certified-general.

There's a major lack of public awareness of who appraisers are, what we do, and why we do it, and I think that's the reason this field is in such demand. I mean, I'd been a real estate agent for *ten years*, and I was really unaware of what the appraisers did. I knew there were people involved in the transaction when I would sell a house who had to come out and "do the appraisal," but beyond that, I really didn't give it much thought.

Throughout that ten years of sales, however, there were multiple times when I "needed an appraiser," and eventually, I got to know a few of them. Because of this, when the time came for needing a mentor, I had connections to leverage. It didn't take long at all before I was introduced to a supervising appraiser who mentored me through successful completion of the three-thousand-hour certification process.

The real estate sector deals with real estate, so it stands to reason that real estate connections can transfer to the real estate side of appraising highway infrastructure projects.

MENTORING

Within the appraisal certification process, education and supervised, on-the-job training, or mentoring, are equal requirements for certification. Opportunities for employment that include supervised-training hours can be obtained by working your way up in a company. Larger appraisal businesses, right-of-way agencies, and acquisition or engineering firms may be places to seek these types of arrangements. Working for an established company may not speak to noncorporate folks, but for some, working for a company that incorporates their mentorship requirements is ideal, and they end up remaining with that company for the remainder of their career.

For the general certified appraiser who prefers working on their own, however, finding that mentor becomes a serious challenge. I've heard horror stories about people finishing their training classes and contacting hundreds of appraisers without having any luck finding a mentor. This really wasn't an obstacle for me because I had real estate industry connections that I was able to leverage. Again, a great starting point for appraisers is in real estate sales.

MENTORS IN THE NICHE SPACES

General appraisers, specifically in the eminent domain sector, seem to be more receptive to having trainees on staff than do residential appraisers who do mortgage work. Because their trainee will be eligible to be their competition shortly after certification, the residential business model tends to lead trainers to perceive training as a loss to their own time and money.

Conversely, in the niche spaces, becoming the supervisor's immediate competition is virtually impossible because most states require three to five years of postcertification experience to get on their approved work lists.

The general appraiser also spends quite a bit of time researching and developing the appraisal. This leaves room for a trainee to take on smaller pieces of the overall project, such as writing up minor components of a report. Many eminent domain appraisers I know keep at least a couple of trainees on at any given time and find it to be a mutually beneficial relationship.

One rule for trainers to keep in mind is that the Appraiser Qualifications Board restricts them from having more than three trainees attached to their license at any time. This is to prevent trainers from having too many trainees to realistically have time to invest in. One drawback of this rule is that limiting trainees also limits the general certified appraiser from growing and scaling a staff.

"The Real Property Appraiser Qualification Criteria (Criteria) restricts supervisory appraisers to a maximum of three trainee appraisers 'at one time' (unless the state has a program that would allow for more under the Criteria)."[79]

CHARACTERISTICS THAT FIT WELL IN APPRAISAL WORK

Whereas residential appraisers might spend half their time in the field, a general appraiser may spend days, even weeks, belly up to a desk, researching anything from comparable sales to rental data to building costs—all the things that go into developing a general appraisal. The depth of analysis required is significant, and much of it gets done at a computer. Whereas real estate sales "success" comes from a personality, people-oriented standpoint, appraisal work is more analytical in nature. There are a lot of moving parts in play with this profession, so the person who can juggle several things at once and maintain accuracy has an advantage. Specifically, six mindsets or characteristics strike me as necessary for success in this field:

1. Analytical

If I were looking for a trainee with their whole career ahead of them, I'd seek out an analytical-type person who thrives on research. People who flow naturally in these traits tend to find appraisal work to be an outstanding career fit.

---

79   Appraisal Foundation, "Appraiser Qualifications Board Q&As," June 2017 (Vol. 9, No. 1), https://appraisalfoundation.org/imis/TAF/AQB_QAs.aspx.

2. Skilled in Persuasive Writing

Two early identifiers for someone geared for appraisal work, in my experience, are talents in persuasive writing and research-paper-type projects. Ease with this writing style reflects not only the proper report tone but also the mindset of the writer.

3. Proficient in Math

Appraisals deal in numbers, so some math proficiency is necessary, but it really doesn't go much beyond a college-algebra level. We're not doing complex calculus on a regular basis or anything like that, but it's important that you have a general comfort level with mathematics.

4. Detail Oriented

An appraiser must be able to see things the average person doesn't even know are there, and they must be able to judge what they see with well-supported opinions. Attention to detail is imperative.

5. Independently Motivated

Independent motivation will be necessary to get the job done, especially for self-employed appraisers. Independent appraisers have to generate all their own business, respond to inquiries in a timely fashion, submit reports by their deadline dates, keep all of their business records, et cetera. There is no boss telling them to get it done, so they must be able to independently maintain all aspects of their business.

6. Comfortable with Solitude

Appraisers can be somewhat reclusive individuals, and they must be comfortable with solitude. More strongly extroverted individuals may find the work grueling, whereas someone with an appraiser's personality will be contently focused on the computer for long periods of time.

Networking and making connections are important, but it's not central to their day, while long hours of research will dominate their time.

THE FUTURE OF THE APPRAISAL FIELD

Generally speaking, there will always be a need for the appraisal process, and certainly, there will always be highway infrastructure of some kind that will need general appraisers. However, over the past thirty-plus years, appraisers have been increasingly concerned with the developments in technology and artificial intelligence. Some have a true fear that AI will replace people in appraisal work, just as travel agents were made virtually extinct by online travel planning via the home computer.

I've tested ChatGPT and Google's Bard by asking a couple of questions. It's been hit or miss as far as staying accurate when the discussion gets complex—as appraisals will. As of right now, I think the best use of AI is not to be terrified of it but to use AI for limited tasks. Appraisers aren't going to send Bard to court, but they might use it to help them write a report more efficiently.

SEVEN STEPS TO BECOMING A GENERAL CERTIFIED APPRAISER

For those formulating a path toward an appraisal career, here are the seven steps that worked for me. Your circumstances may require individual adjustments, but this will give you an idea of what to plan for. Remember to allow yourself four or more years to complete all seven steps.

1. Starting as a real estate agent was key to accelerating my path. This step provided exposure to the overall real estate process. Again, this can provide trainees the income, flexibility, and needed connections to conquer each successive step. (Earn while you learn!)

2. Earn your bachelor's degree.

3. Take the first seventy-five hours of supervisory appraiser/ trainee courses.

4. Find your mentor.

5. Gain the Licensed Residential Appraiser credential. This will require an additional 75 hours of training, for a total of 150 hours, plus 1,000 hours of in-field supervised training. This begins the supervised coursework the trainee's mentor will sign off on. Most states have a formal process for the supervisor to log training hours.

6. Continue moving forward by earning the Certified Residential Appraiser credential. (Technically, this is not a requirement for general certification, but it offers a valuable layer of experience.) The CRA requires an additional 50 hours of coursework, for a total of 200 hours, plus 1,500 supervised, in-field hours.

7. The Certified General Appraiser license requires an additional 100 hours of coursework, for a total of 300 hours, plus 3,000 supervised, in-field hours. Of these hours, 1,500 must be nonresidential work.

• • •

Good appraisers will regularly tap into their critical thinking and analysis-and-research skills. We get paid to solve puzzles for a living in this profession. If that excites you, it's possible you'll do well as an appraiser.

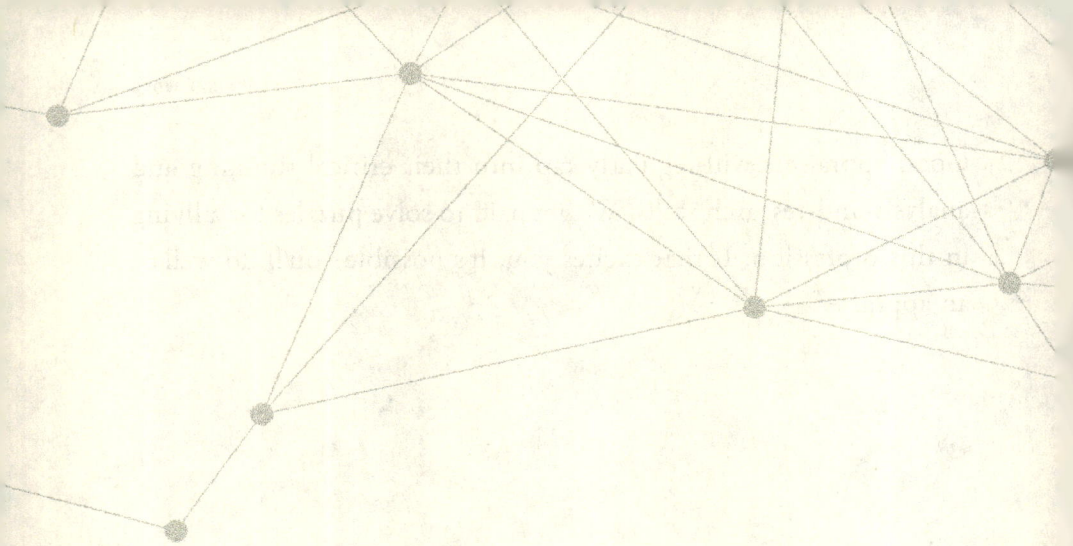

10

TITLE SPECIALISTS

Career Information

PROVIDED BY ESTHER FRANKLIN

Esther M. Franklin is a C-level real estate consultant with extensive title expertise and a boundless desire to drive her industry forward. Over the course of her nineteen-year tenure in the real estate sector, she has earned a bachelor of arts in legal studies from the University of Pittsburgh, an MBA from Colorado Technical University, and numerous certifications and educational credits from other institutions, to include Duquesne University Paralegal Institute, New York University Schack Institute of Real Estate, and the Tuck School of Business at Dartmouth.

Esther is not only both a published author and licensed real estate title agent in seven states, but she is also the chief executive officer of ALROWS, LLC and Tri-State Paralegal Service. Esther's skill set extends to legal project management, business development, content writing, marketing, and corporate communications. She approaches her work with passion, placing particular emphasis on implementing significant initiatives, building strong teams, establishing trust, ensuring safe and synergistic working environments, and sustainably increasing revenue.

About the Career

Definition of Title Agent (Bureau of Labor Statistics): "Search real estate records, examine titles, or summarize pertinent legal or insurance documents or details for a variety of purposes. May compile lists of mortgages, contracts, and other instruments pertaining to titles by searching public and private records for law firms, real estate agencies, or title insurance companies. Excludes 'Loan Officers.'"[80]

National Average Salary (Salary.com): $53,918. Range: $49,243 to $60,822 annually.

Current State of Title Work in Infrastructure: Needing immediate candidates.

Opportunities to Work for a Company? As an abstractor, there are opportunities to work within an organization as an employee; title companies are always looking for abstractors to get their searches done. Working in an organization comes with a strict adherence to a work schedule and company policies/procedure. Abstractors will either be paid as hourly employees or salaried, depending on their level of expertise.

Opportunities for Self-Employment? Yes; title specialists can opt out and become self-employed by working as subcontractors. Self-employed abstractors are responsible for securing their own workload by marketing their services. This can become cumbersome; however, it can be well worth it if they are diligent. In addition, self-employed title specialists must be responsible for errors and omissions insurance, fidelity bonds, surety bonds, general liability insurance, and filing their quarterly and year-end taxes. Keeping a good accounting system to track income and expenses is critical. An additional expense to consider is software; some clients may require certain software systems to make the workflow process seamless.

THE TITLE SPECIALIST CAREER

The term *title* describes the right of a person or group to own a legally defined piece of property, most often a residence, business property, or land parcel. To earn this legal right, the owner of the property must be in possession of documents that accurately describe a property's ownership and history.[81] These legal documents are referred to as the title.

---

80   U.S. Bureau of Labor Statistics, "Title Examiners, Abstractors, and Searchers," updated April 3, 2024, https://www.bls.gov/oes/2023/may/oes232093.htm.

81   Rob Stewart, "What Is a Title Agent? & How to Choose the Right Title Agent," LinkedIn, accessed September 12, 2023, https://www.linkedin.com/pulse/what-title-agent-how-choose-right-rob-stewart.

When the sale of a property takes place, whether it be commercial property, residential, or land, the title specialists find out who owns it and how much each interested party is legally authorized to receive for its purchase. To accomplish this, title specialists are responsible for conducting thorough research and examination of property titles. This involves identifying the lawful owners of the properties, checking for any existing liens or encumbrances, and ensuring that all necessary rights and permissions are properly obtained.

Specifically for the infrastructure rebuild, title specialists will focus on preventing potential legal disputes or delays that can arise from issues related to property ownership rights. They provide expert guidance and assistance throughout the project, helping to ensure its successful completion while protecting the interests of both the public and the property owners involved.

Title specialist is a broad term that can really be broken down into two categories: the abstractor and the title agent. Both are critical to the overall process.

THE ABSTRACTOR

"What is an abstract of title? In essence, the abstract of title is a chronological document that summarizes everything that has happened with the title of the property. It starts from the time the property was first recorded as owned and continues all the way to present day. The document tracks every transaction on the property, starting with the initial grant deed and followed by a record of every instance in which the property title changes hands. It also includes a record of other key details, like liens against the property."[82]

The abstract is an actual report developed by the abstractor, or the person who performs the search. Each abstractor or title agent is responsible to know and follow the rules in the state where they are performing their research.

A noncomplex search can be done online, as following the chain of title is manageable. Depending on the nature of the request, the person examining the title will often put the information in a comprehensive report that notes the legal ownership as far back as sixty years or more.

A more complex abstract, for something like a sewer project or cell tower search, will require that the abstractors go to the recorder of deeds office for their search. Some online records don't go as far back as needed in these searches, so the abstractors will physically search through the land records index information to find the history of landownership.

For the title agent specializing in highway infrastructure builds, the project starts when the engineer sends us a right-of-way plan stating the number of acquisitions they need. So let's say Montgomery County, Pennsylvania, is building a park to create green space available to a minority community. There is an ideal location along the highway. The first step in acquiring that land is to find out the true ownership of that "ideal location." The abstract or title search is basically the first part of the acquisition process. You're learning who owns the interest in the property.

82 Gracie Goff, "What Is an Abstract of Title?," Bankrate.com, October 17, 2022, https://www.bankrate.com/real-estate/abstract-of-title/#:~:text=In%20essence%2C%20the%20abstract%20of,the%20way%20to%20present%20day.

Applying all information in each component of the title process can get complex; however, there are basically six steps to what will need to be accomplished for the complete abstract:

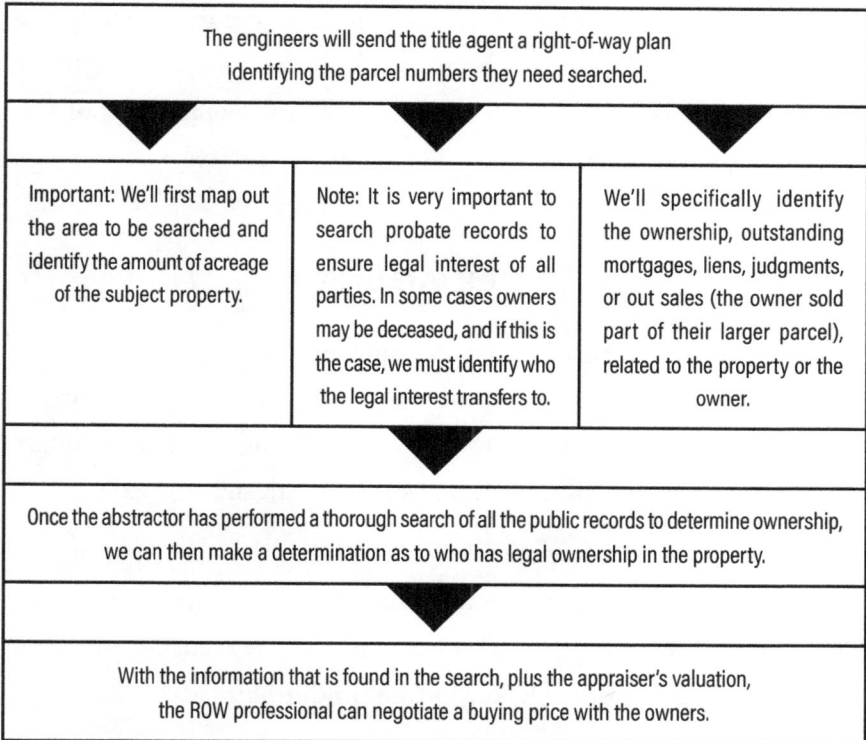

The engineers will send the title agent a right-of-way plan identifying the parcel numbers they need searched.		
Important: We'll first map out the area to be searched and identify the amount of acreage of the subject property.	Note: It is very important to search probate records to ensure legal interest of all parties. In some cases owners may be deceased, and if this is the case, we must identify who the legal interest transfers to.	We'll specifically identify the ownership, outstanding mortgages, liens, judgments, or out sales (the owner sold part of their larger parcel), related to the property or the owner.
Once the abstractor has performed a thorough search of all the public records to determine ownership, we can then make a determination as to who has legal ownership in the property.		
With the information that is found in the search, plus the appraiser's valuation, the ROW professional can negotiate a buying price with the owners.		

THE SEARCH IN ACTION

By way of an example of how the search works in everyday property sales, a family I know was buying a home from a private owner. The owner had lived there all his life, and he and his family had taken sole possession of the house more than twenty years earlier. There wasn't even a question as to who owned the property. However, when the title search was conducted—meaning the abstractor went to the recorder of deeds' office and physically looked through the records, the owner's sister was listed as a co-owner.

Turns out, the property was willed to the *two* siblings by their late father. The owner's sister had to sign off on the deed transfer, conveying her half interest to her brother (clear the title), before the sale could be finalized. If the homeowner's sister would have refused to convey her half interest, she would have still owned half of the property, even though she hadn't lived there in forty years.

TITLE AGENT

Once the abstractor has collected the search information and developed the abstract, the process goes into a second phase, which requires a licensed title agent. Rob Stewart, real estate closing expert with Smart Secure Real Estate Closings, explains what a title agent does:

> "A title agent is responsible for certifying the validity of a title on a piece of real estate, which includes guaranteeing proper ownership of a clean title as well as securing title insurance to protect the buyer from undue harm after a sale.

> "Title agents typically work alongside other members of a buyer or seller's real estate team. This might include the real estate agent, certified lender, closing attorney, and other individuals involved in the sale. In locations where the title agent is responsible for ensuring that the actual transaction happens smoothly, these professionals work one-on-one with banks and lenders.

"A title agent can help you navigate both the legalities of property ownership as well as the organization of important paperwork that must be on file when you buy or sell."[83]

To illustrate what a title agent does, let's return to the Montgomery County, Pennsylvania, park from our earlier example. If a lender is going to lend the town parks commission several hundred thousand dollars on the construction of a new park, the title agent's job is to ensure that no one can take that property by claiming legal ownership. Title agents verify that it's safe for the lender to loan the money to the buyer. If the title agent finds ownership to be other than the parties who are claiming interest, they'd have to ensure clear title, either by way of order from the court or by having the rightful owners convey interest via deed.

While abstractors research the land to determine who holds legal interest in it, title agents perform the settlement and closing of the sale of the property. This service includes preparing settlement statements; obtaining tax certifications, documents, and deeds; preparing the mortgage; and collecting money for the transfer and distribution of funds to the buyer and seller at closing.

EDUCATION AND REQUIREMENTS

In truth, all title agents are abstractors, but not all abstractors are title agents. When it comes right down to it, the difference between abstractors and agents is the licensing that the title agent acquires. Title agents must be licensed in the state where they reside. For this license, there is a slightly different set of educational requirements than for that of the abstractor. Take a look at the side-by-side com-

---

83   Rob Stewart, "What Is a Title Agent?"

parison below. Of particular note are the more-involved requirements of the title agent:

Abstractor Education Requirements:

"The absolute minimum education to be an abstractor is a high school diploma or GED, but you are more likely to find a position if you have an associate's or bachelor's degree from an accredited college or university. A degree in paralegal studies or business is helpful.

"Certification is available through the National Association of Land Title Examiners and Abstractors (NALTEA), but it isn't a requirement. It is one way to show that a certifying organization has vetted your knowledge and requires you to keep up in your field through continuing education.

"Another organization, the American Land Title Association (ALTA), has a three-month course you can take to develop a foundation of knowledge about abstracting. It includes methods and procedures for title examination; topics on title insurance and government agencies; and tips about leadership and management. At the end, you receive a certificate of achievement."[84]

Title Agent Education Requirements:

"How to become a title agent: The requirements for becoming a title agent vary from state to state. However, you can follow these basic steps to become a title agent:

Earn at least a high school diploma or GED: Most title agents need to be at least 18 years old and have a high school diploma or equivalent. You could also consider earning an associate's or bachelor's degree in a field like business, finance, marketing, or accounting to increase your employability. Search for title agent jobs in your area to see if employers commonly seek higher education in a specific field to help you prepare for your next steps.

Complete a title agent course: Completing a pre-licensing course is a requirement to become a title agent, and each state has its own course standards and guidelines. The course length varies by state, but could be at least 15 hours and up to 40. The courses cover real estate laws, including proper selling and buying practices, and insurance. Learning about these real estate practices can ensure you provide the most comprehensive services for buyers/sellers.

---

84   Courthouse Direct, "How to Become an Abstractor: Required Education & Skills," accessed September 12, 2023, https://info.courthousedirect.com/blog/how-to-become-an-abstractor.

Title Agent Education Requirements:

Pass an exam: All title agents need to pass a licensing exam to provide their services in the real estate industry. The exams usually cover the topics you learn in the licensing course, including local, state, and national real estate laws, title searches, and basic mortgage procedures. You likely need to take the computer-based exam at a designated testing site in your area.

Keep your license current: Real estate laws and regulations continually change, so title agents need to complete continuing education to maintain the most up-to-date knowledge. You may need to complete a certain number of continuing education hours and submit proof to your state's licensing board. Continuing education can include real estate seminars and title agent classes, where instructors teach you about best real estate practices. You can check with your state's board to learn more about the continuing education requirements."[85]

In the real world, if I'm hiring either an abstractor or an agent, their degree or diploma really isn't as significant as their work experience and drive. For instance, if an abstractor or agent applies with a high school diploma as their highest level of education but they've worked for federal agencies, that will offset any preconceived ideas I had about requiring a degree. Why? Because anyone who's worked at federal agencies is likely to have some tough skin, and there were certain milestones they had to achieve.

On the other hand, if the candidate has a college degree but they work at Starbucks, the necessary drive for the job seems lacking. I will have to consider how long it will take them to get up to speed. So a degree doesn't pave the way right into this field, and neither does a lack of higher education keep someone from success in this industry.

Involvement with established associations and networking platforms, however, may streamline a title specialist's efforts. Not only can association involvement connect them with jobs, but some continuing education is frequently offered through participation.

85   Indeed Editorial Team, "How to Become a Title Agent in 4 Steps," Indeed Career Guide, January 26, 2023, https://www.indeed.com/career-advice/finding-a-job/how-to-become-title-agent.

Getting involved with relevant associations also presents chances for those needing jobs to connect with those offering jobs—and improve their résumés for future jobs at the same time. Listed below are a few organizations that I follow that may be beneficial to education and resource seekers:

Associations/Platforms Beneficial to Title Specialists/Candidates	
American Land Title Association	www.alta.org
American Society of Highway Engineers (ASHE)	www.ashe.pro
Appraisal Network, Corp	www.appraisalnetworkcorp.com
Construction Management Association of America (CMAA)	www.cmaanet.org
Mortgage Bankers Association	www.mba.org
National Association of Land Title Examiners and Abstractors (NALTEA)	www.naltea.org

MENTORING REQUIREMENTS

There are no specific mentoring or supervised hours required to become a title specialist; however, it's nearly impossible to learn and move forward in the field without one. My first work experience in the real estate sector was in 2004. It was with West Penn Financial, and it was located in the Strip District. I received that job through a temporary placement agency, and I am thankful for that opportunity. Coming from a legal background, I could clearly see that the legal profession and the real estate sector coincided. In fact, it seemed to be the perfect fit for me, and at that point, I found my passion.

My experience came from working with various mortgage companies over the course of many years. To place myself in line for higher positions within the industry, I diligently educated myself on title insurance requirements, got involved in organizations, and went after the education requirements.

The mentor who introduced me to the right-of-way sector was Fred Brient, from Orion Land Services. Fred found my company via a Google search; he was looking for a subcontractor to team with on a highway project for the Pennsylvania Turnpike Commission. Talk about learning to swim in the deep end of the pool; I was oblivious, prior to this meeting, that a title search related to transportation in any way.

My business-degree courses never even alluded to some of the processes that were standard in property title and right-of-way work. It was like starting over again with something completely foreign. I was eager to tackle it, and interested abstractor/agent candidates must approach the industry in the same way. If they're going to panic and run, this is not the industry for them. If they'll face challenges head-on until they're conquered, they will likely succeed. The time working with their mentor allows them to grow to this critical point of confidence.

CHARACTERISTICS THAT FIT WELL IN TITLE WORK

The detail-oriented person who enjoys research tends to do best at title work. These individuals appreciate a challenge, and they have the mental energy to stay on top of several complex components that are all relevant at the same time. This candidate must be comfortable both in front of a computer, writing up the abstract report, and in their car, traveling to courthouses for information research. Paralegals or individuals with proven research and organizational skills shift most easily into title work. Military personnel, in my estimation, also show a particular discipline that seems to transition well in this field.

Listed below is the job description we actually use when seeking title candidates. If these duties sound interesting, even career fulfilling, that candidate is likely a good fit for the job. If the list sounds boring, even torturous, that individual typically will not thrive in title work.

Title Specialist Job Description

- Search computerized and manual title information property records and determine instruments filed in public records. May map legal descriptions to obtain correct search parameters.

- Determine and ensure that property has been researched properly and information relating to parties involved and taxes is accurate. Ensure that base title information is within the Fund's guidelines.

- Examine all documents in chain of title for validity based on comprehensive knowledge of legal descriptions, real property laws, and the Fund's Title Notes.

- Organize title information; ensure proper documentation of research evidence.

- Prepare final product utilizing internal programs, or may prepare worksheet from which final product will be prepared. Attach copies of applicable instruments and ad valorem tax information.

- Progressively advance examination knowledge and experience by supporting increasingly complex title transactions under direct supervision of a senior examiner or manager.

- Interact with internal and external customers to discuss findings, documentation needed, and answer inquiries. Forward escalated issues or inquiries to the manager as necessary.

- Identify and communicate with attorney any issues relating to title. Provide assistance with resolving issues.

- Participate in quality assurance review of examined products prior to delivery to customer as required.

- Assist in training and coaching of examiner trainees.

- Maintain current awareness and knowledge of changes in laws, regulations, and procedures affecting real property.

- Knowledge of the title insurance and information industry is preferred.

- High school diploma or equivalent required.

- Prior title examination experience strongly preferred.

Skills:

- Oral and written communication skills

- Detail oriented

- Problem-solving

- Planning and organizational skills

- Math proficiency

- Windows and Microsoft Office

THE FUTURE OF THE TITLE FIELD

I see the future of the title field being busy, productive, and steadily evolving! The future of title research is likely to be heavily influenced by advancements in technology. There is great potential for increased efficiency, accuracy, and transparency through the integration of artificial intelligence and emerging technologies.

While AI systems can offer significant benefits to title research, it's important to be mindful of their limitations and potential ethical considerations. A balance between AI and human involvement can help harness the advantages of something so new while maintaining accountability and addressing potential drawbacks, such as privacy and security.

As imposing as AI is expected to be, however, I don't believe it threatens the title industry as much as the absence of qualified specialists will. I firmly believe that finding qualified abstractors who can competently oversee the research needed for title work is the biggest challenge to come. Just as the baby boomer retirements are affecting government departments everywhere, the title industry is also seeing record numbers of retirements.

The BIL's provisions for new and expanding infrastructure may cause thousands of acquisitions to flood the title field in relatively short bursts of time. Each and every one will need the title searched and an agent to finalize it. This field is in serious need of determined, qualified individuals to become first-rate title specialists! The BIL provides for many highway projects to take place over a long period of time, so you can feel confident in the future of the field.

PART 4

HOW THE FIVE CAREERS NEEDING IMMEDIATE FULFILLMENT INTERCONNECT

11

HOW THE FIVE CAREERS NEEDING IMMEDIATE FULFILLMENT INTERCONNECT

To illustrate how our five careers needing immediate fulfillment work together, we'll say that the community of Lakeside, in the city of Buffalo, New York, qualifies for several programs within the BIL's provisions. The neighborhood is predominantly minority occupied; smog is a constant health concern for these residents, as fumes from traffic regularly hover over their atmosphere; and shopping and medical services require the seeker to travel straight into the congestion of downtown or ten miles out of their way.

To remedy this, the Greater Buffalo Niagara Regional Transportation Council (the agency) has decided to use the BIL's available funding to build a two-mile stretch of road that will connect several neighborhoods with essential business and medical services. The site they've selected is relatively wide open with just a few businesses and homes on Easton Street, along the outskirts of the route. The path will directly connect to a thriving plaza just outside the city.

To begin this endeavor, the GBNRTC engages a full-service engineer and architecture firm to manage the project. An engineering-architect (E/A) team is involved early on to gain an understanding of the area from an engineering and design standpoint. The predevelopment due diligence is where the engineers concentrate their expertise.

The E/A team combs through old maps and geographic-information systems to see what information is publicly available about the agency's preferred route. In doing so, they discover a few problems with the desired location for the road.

For one, the agency's plan cuts through one home that was built in 1942. Asbestos is a big problem when demolishing buildings of this era, and it will take time, money, and expertise to abate it. Across the street from this home is a historical art museum, and there are laws against tearing it down. Worse, some of the area on the east foundation site has archaeologically sensitive American Indian remnants beneath the soil, and there are established statutes prohibiting excavation in the area.

The E/A team goes to work surveying the land for options that will avoid these problem areas. If they add a curve in the road at the most central point and move the ramp foundation site fifty yards to the west, there is nothing but open fields. The biggest problem, however, is that part of the area is a complete wetland. The Environmental Protection Agency (EPA) will not permit any structure to be built on such unstable soil.

The E/A team quickly determines that a bridge can span over the wetlands. The architects take the lead now to create a design for this change in plan. Their design adjustments will increase the budget by about 20 percent, so they present their findings to the agency, GBNRTC. A meeting is scheduled for next Wednesday, where the team explains their solutions to the previous plan's challenges. Nothing is decided at the conclusion of their presentation; it takes time for the agency to review the changes and approve the plan.

Six months later, and after several meetings, calls, submissions, and sessions of addressing review comments, the E/A team receives notification that the plan is approved. They get to work, collaborating and laboring through details to ensure the bridge is beautifully and practically designed as well as structurally sound. To do this, they intentionally question all aspects of the bridge's end use:

What types of vehicles will be traveling over it? Are we moving toward a more public transportation route? What type of bridge should be constructed? Would a suspension bridge be appropriate? How much will it weigh? What are the various loads, especially wind and snow loads, since Buffalo is known for blizzards? Is this an opportunity for below-grade transportation, not necessarily above grade?

> "Grade is the rate of change of the vertical alignment. Grade affects vehicle speed and vehicle control, particularly for large trucks. The adopted criteria express values for both maximum and minimum grade."[86]

The foundation engineer steps in to determine if taking the load down to a deeper stratum will accommodate the design. A gas line is right in the path now, so that must be relocated safely, with minimal disruption to any residents.

Three years later, after all the considerations have been investigated and the environmental engineers have performed endless examinations and testing of the site, a plan is created that highlights any problem areas. To actually get that plan to the point of construction,

86   "Safety, Grade, Highway," US Department of Transportation Federal Administration Archives, accessed September 12, 2023, https://safety.fhwa.dot.gov/geometric/pubs/mitigationstrategies/chapter3/3_grade. cfm#:~:text=Grade%20is%20the%20rate%20of,both%20maximum%20and%20minimum%20grade.

they need to know who owns the parcels along Easton Street that they will need in order to complete the route.

Enter the ROW professionals and title agents.

• • •

When the plan becomes available to the ROW firm, the first step is to assign a ROW professional to the job. Mr. Smith will take the lead on the Easton Street Project, so his first course of action is to visit the site and develop a field report, explaining his own observations of the current landscape.

The design is fairly straightforward. Since this particular project began three-plus years ago, two additional businesses have gone up along that route—and of course, the new roadway will impact them. Mr. Smith can see where the curve in the road avoided some difficult soil, but guardrails, which are required as the ramp progresses, would shave a three-foot sliver of frontage off these two businesses. One residential home on Easton Street would also require a three-by-forty-foot section off the front yard, and the final residence, a ranch with a well-cared-for picket fence, would require complete demolition, as it sits in the direct path of the ramp's initial incline.

Mr. Smith records his observations on the field report and returns the plan to the engineers, who make adjustments based on the updated information. Any changes travel back through a pipeline of checks and balances before the plan is finally returned to Mr. Smith.

The abstractors step in at this point, either from the ROW agency or title agency—both do abstract work. In our case, a title agent, Ms. Braun, is retained, and her team completes the abstract. Since this is a highway project, not a simple title search, the abstractors physically go to the city courthouse to verify landownership. They locate relevant phone numbers and contact information, pull titles, review deeds, and identify the amount of acreage on each property.

The abstractors find the affected landowners, whose properties fall under eminent domain law. Melanie's Hair Salon, Reliable Bicycle Repair Shop, and homeowner Mrs. Krell are those requiring three-by-forty-foot sections off their frontage. The Dean family's complete residence is needed.

Mr. Smith calls each landowner personally to alert them to the project and let them know that their property lies in its path. This is not an official call; it's more of an icebreaker call, meant to alert them to the circumstances and allow them time to process it. Mr. Smith emphasizes that an official letter will be coming in the mail and that it is important for them to read the letter and follow any instructions.

The ROW and title teams wait several months for the agency to approve all the plans; then finally, they receive the notice to proceed. At this time, Mr. Smith releases official notices of eminent domain to the landowners. He requests sit-down meetings with them, and he fully explains how their properties will be affected. He answers any questions they may have and asks them to read several pamphlets further explaining the details, and the homeowner completes a W-9 that will be needed later in the process.

At about the same time that Mr. Smith notifies landowners of their impending circumstances, eminent domain appraiser Mr. Rudd also receives the plan from the agency. The appraisal is central to ROW professional Smith's process for assessing fair compensation for the landowners.

Appraiser Rudd's first course of action is to check the scope of the project to see what the impact of the roadway will be on the overall area. In this case, one property is being appraised for its total value, which will require an in-depth appraisal. The other three properties only need to be appraised for the loss of three-by-forty-foot sections from their frontages, which is less complex. After Rudd reviews the property and prepares valuations, he writes his report and returns it to the client.

Upon approval, a signed copy of the appraisal is sent to ROW professional Smith, who will develop monetary offers. A review team from Mr. Smith's company steps in to inspect all documents before sending them to the respective parties. If the paperwork is not perfect, it could further hold up the process, so the project manager and support team are meticulous.

At this point, offers are presented to each respective landowner. Melanie's Hair Salon is barely affected, as the frontage comes off a grassy area in front of her parking lot. She's overjoyed by the offer and signs the land over directly.

Rick's Bicycle Repair Shop isn't as fortunate. Three feet off their frontage cuts into a row of parking from their storefront. They aren't happy at all; the only way to retain sufficient parking for their customers is to reconfigure their parking area. To accomplish that, they must close the shop completely for at least one week! Mr. Smith steps in to assure the owner that these considerations have been amply provided for in his offer. When Rick still hesitates, Mr. Smith reminds him that this project will benefit the community and cause an influx of new passersby who may need bicycle service. Rick considers this and finally settles amicably.

The homeowners aren't as easy to deal with. Mr. and Mrs. Dean are confronted with the loss of their entire property. They have a family; they've worked hard to build what they own; it is shocking to learn that the law of the land is against them. In this situation, the agency will relocate them, but current real estate prices versus the value of the Dean residence are at completely different price points. They want to be relocated very near the business park so the children can attend a good school. Negotiations become quite involved before they finally find something adequate. Nearly a year later, the family settles in to their new home. Thankfully, they're pleased with the bustling new location, which is only one block from top-rated elementary and junior high schools.

The fourth acquisition is the most disruptive. Mrs. Krell is adamant that if three feet of her front yard are removed "to accommodate an ugly guardrail," it compromises the entire front of her home. She wants to know the actual law on guardrail placement, so Mr. Smith shows her the local municipal codes of the city. Mrs. Krell doubles down and absolutely refuses to sell unless she is compensated for the full value of her home in exchange for the needed three-by-forty-foot parcel.

Mr. Smith stands just as firmly and explains that she is really only harming herself with that decision because her property will be condemned, meaning the agency will acquire her home by force rather than fair compensation if she chooses to take it to that level. Mrs. Krell finally, begrudgingly, agrees to the terms. She narrowly avoids condemnation, but her objections hold progress up for an additional seven weeks.

• • •

As she gets into this case, the first thing title agent Braun does is conduct a thorough title search and examination to establish clear titles for the Easton Street acquisitions. She checks for any outstanding mortgages, unpaid taxes, judgments, or other issues that may affect the agency's ownership rights.

As part of her due diligence, Ms. Braun will contact all the taxing bodies in the jurisdiction of the property to obtain tax certifications and "no-lien letters" to ensure all taxes have been paid on the property being acquired.

> **"No-lien letter** means a letter which makes clear that a third-party sub-custodian's right to claim a lien and/or right of retention and/or sale over the assets in an account is restricted."[87]

It's during her search that Ms. Braun discovers an outstanding lien against Mr. Dean's property. Having a clear title is crucial, as it ensures that the property can be transferred to the agency without any legal complications or claims from third parties. The Deans' mortgage is considered a lien or an encumbrance on the property.

As the title agent, Ms. Braun must contact the bank that holds the mortgage to the property to obtain a payoff statement that will state the amount payable to the bank to clear the lien. The cost associated with the purchase of the Dean property will include paying off the outstanding debt. Any funds remaining will go to the Deans. The relocation assistance is a separate transaction; however, they can decide to use the remaining funds in conjunction with the relocation assistance to acquire their new residence.

Title Agent Braun's license authorizes her to perform real estate transactions, so at this point, she prepares the settlement statement, which outlines all the costs associated with the transaction. Once both parties finalize the numbers, the GBNRTC (agency) wires the funds to Ms. Braun's company. She then schedules the closing to take place.

---

87   Law Insider, s.v., "No-Lien Letter," accessed September 12, 2023, https://www.lawinsider.com/dictionary/no-lien-letter.

The buyer (the agency), the seller (Mr. Dean), and the title agent (Ms. Braun)—or a rep from her agency—will be present at closing.

Ms. Braun disburses funds to the appropriate parties, including the payoff of the loan, in accordance with a HUD settlement statement. This worksheet is utilized by title agents to provide a transparent picture of how funds are to be disbursed at closing.

> **"What is a HUD-1 Settlement Statement?** The HUD-1 Settlement Statement is a document that lists all charges and credits to the buyer and to the seller in a real estate settlement, or all the charges in a mortgage refinance."[88]

Once these funds are disbursed, all liens are paid, and the deed and/or mortgage is recorded at the recorder of deeds' office, the title agent can then issue a title policy for this transaction, *and the project can now move into the construction phase.*

• • •

Now, let's look at the very same project when just one career field does not have sufficient personnel. In this scenario, as in the last, the Easton Street plan has been in production for three years; it's gone through all the ecological studies and verifications, and the ROW agent has it in hand. He's finished his assessment and returned the updates to the engineers, and then—slam!—there isn't an available eminent domain appraiser to keep the project moving forward. Progress comes to a screeching halt because this job specifically needs an eminent domain appraiser.

A chain reaction occurs. The job is halted; the respective staff, of course, move to the next pressing project; an appraiser is needed on

---

88   Consumer Financial Protection Bureau, "What Is a HUD-1 Settlement Statement?," accessed September 12, 2023, https://www.consumerfinance.gov/ask-cfpb/what-is-a-hud-1-settlement-statement-en-178.

that job, too, so it's added to the pile; the completion of all projects slows down to a crawl due to the backup; work that you've already done cannot be submitted without that eminent domain appraisal, so payment also lags behind.

Files will sit on desks for months in some cases, piling high and collecting dust, when they hit the real estate sector of the project and competent ROW professionals, appraisers, and title agents can't be found. Worse, the community in serious need of the healthcare and business opportunities this roadway will provide goes through yet another harsh winter without any progress being made.

PART 5

CAREERS RELATED TO THE INFRASTRUCTURE REBUILD THAT ARE EXPECTED TO BE IN DEMAND

12

CAREERS RELATED TO THE INFRASTRUCTURE REBUILD THAT ARE EXPECTED TO BE IN DEMAND

Beyond the five highlighted careers-in-need, there are scores of jobs that are related to infrastructure that may be in demand as well. As we've already discussed, jobs in seemingly unrelated fields tend to benefit from building projects too. Doctors, lawyers, accountants, writers, sales experts, and office personnel of all kinds will be needed all over the nation as these builds progress.

At the same time, minority business owners should keep in mind the four programs available to you that are highlighted in chapter 3: Affirmative Action, SBA 8(a), the Reconnecting Communities Program, and the Disadvantaged Business Enterprises. Now is the time to get that degree, license, or certification in your field of interest.

Join these programs to get everything out of your business that you can in order to effectively serve your community.

Below are eight additional professional arenas expected to closely follow on the heels of the infrastructure rebuild. To help readers on their journey, we've provided an overview of each field and a comprehensive list of jobs you can look into if the field piques your interest.

Eight Fields Expected to Be in Demand Along with the Rebuild

1	Broadband	Page 170
2	Autonomous Vehicles	Page 173
3	Renewable Energy	Page 176
4	Information Technology	Page 179
5	GPS/GIS	Page 183
6	Drones	Page 189
7	Cryptocurrency	Page 192
8	Alternate Modes of Transportation	Page 197

BROADBAND

"Broadband is the transmission of wide bandwidth data over a high-speed internet connection ... According to the FCC, the definition of broadband internet is a minimum of 25 Mbps download and 3 Mbps upload speeds. Broadband provides high speed internet access via multiple types of technologies including fiber optics, wireless, cable, DSL, and satellite."[89]

Increasing broadband capabilities to rural and underserved areas is a significant part of the BIL. According to Whitehouse.gov, "The Affordable Connectivity Program is the nation's largest broadband affordability program. Created through the Bipartisan Infrastructure Law, it is a $14.2 billion successor program to the Emergency Broadband Benefit."[90]

In other words, broadband experts are needed to bring this critical service to multitudes who have been left behind. Businesses that take advantage of the BIL's provisions and have broadband expertise are poised to reap some of the largest benefits of this infrastructure rebuild.

Jobs in the Broadband Field

Job Title	National Average Salary	Job Description
Broadband Installer [Technician]	$40,000–$42,000/yr[91]	A broadband technician is responsible for the installation, maintenance, and repair of telecommunication systems, including telephone services, broadband internet, and cable television. They regularly perform these services at different businesses and sites. A degree in telecommunications or a related field is highly beneficial, as well as certification and training from a professional body for telecommunication installers.[92]

89   Verizon.com, "Broadband," February 21, 2023, accessed September 12, 2023, https://www.verizon.com/articles/internet-essentials/broadband-definition/.

90   White House Fact Sheet, Briefing Room Statements and Releases, Whitehouse.gov, February 14, 2022, accessed September 12, 2023, https://www.whitehouse.gov/briefing-room/statements-releases/2022/02/14/fact-sheet-biden-harris-administration-announces-10-million-households-enroll-in-broadband-affordability-program-thanks-to-bipartisan-infrastructure-law.

91   ZipRecruiter, "10 of the Most Popular Types of Broadband TV Jobs in 2023," accessed September 12, 2023, https://www.ziprecruiter.com/t/Most-Popular-Types-Of-Broadband-TV-Jobs.

92   "What is Broadband Technician, Job Description and Salary?," Field Engineer, accessed September 12, 2023, https://www.fieldengineer.com/skills/broadband-technician#:~:text=A%20broadband%20technician%20is%20someone,at%20different%20businesses%20and%20sites.

Jobs in the Broadband Field

Job Title	National Average Salary	Job Description
Broadband Engineer	$74,500–$100,500/yr[93]	The duties of a broadband engineer are to install, maintain, and repair a telecommunications network. They are often responsible for ensuring the telecommunications connection is functioning correctly, creating user accounts, and monitoring network activity. Other responsibilities of a broadband engineer are to compile, code, and verify collected data. The qualifications to become a broadband engineer often include a bachelor's degree in telecommunications or a related field and experience in network or software development.
Broadband Sales	$58,000–$93,500/yr[94]	Internet sales representatives are primarily responsible for selling the products and services of the companies they work for online. They are tasked with attaining a specific sales target and ensuring that they are happy and contented with the sales. They also work on the latest online marketing techniques and trends to successfully run their business.
Broadband [Network] Provisioning	$51,000–$78,000/yr[95]	Network provisioning is the process of setting up a network so that authorized users, devices, and servers can access it. In practice, network provisioning primarily concerns connectivity and security, which means a heavy focus on device and identity management.[96]
Broadband Technician	$38,000–$51,000/yr[97]	A broadband technician is someone who is responsible for the installation, maintenance, and repair of telecommunication systems, including telephone services, broadband internet, and cable television. They regularly perform these services at different sites. A degree in telecommunications or a related field is highly beneficial, as well as certification and training from a professional body for telecommunication installers.[98]

93   ZipRecruiter, "10 of the Most Popular Types of Broadband TV Jobs."

94   ZipRecruiter, "10 of the Most Popular Types of Broadband TV Jobs."

95   ZipRecruiter, "10 of the Most Popular Types of Broadband TV Jobs."

96   Cisco, "What Is Network Provisioning?," accessed September 12, 2023, https://www.cisco.com/c/en/us/solutions/automation/what-is-network-provisioning.html.

97   ZipRecruiter, "10 of the Most Popular Types of Broadband TV Jobs."

98   Field Engineer, "What Is a Broadband Technician, Job Description and Salary?," accessed September 12, 2023, https://www.fieldengineer.com/skills/broadband-technician.

Jobs in the Broadband Field

Job Title	National Average Salary	Job Description
Broadband Consultant	$35,500–$83,000/yr[99]	The broadband consultant advises clients on fiber network designs and costs, performs financial analysis and modeling of transactions and financial scenarios, aids development of high-level designs and cost estimate factors, researches and drafts sections of client deliverables, evaluates bids on behalf of clients, and contributes to proposal writing and fielding requests from public agencies, private infrastructure developers, and other potential clients. Broadband consultants also engage in market and regulatory research to distill elements that may impact proposed telecom infrastructure projects and help provide guidance and recommendations for clients to advance specific goals/objectives/programs in an efficient and effective manner.[100]
Broadband Digital Installer	$37,500–$59,000/yr[101]	Installing and/or repairing cable TV and high-speed data or digital services inside and outside of residential and commercial customers' homes and businesses. Performs upgrades, downgrades, identifying technical issues while troubleshooting equipment, measuring, recording, and determining cable signal levels.[102]

Data based on reports accessed September 2023.

AUTONOMOUS VEHICLES

According to Synopsys, an electronic design automation company, "An autonomous car is a vehicle capable of sensing its environment and operating without human involvement. A human passenger is not required to take control of the vehicle at any time, nor is a human passenger required to be present in the vehicle at all. An autonomous

99   ZipRecruiter, "10 of the Most Popular Types of Broadband TV Jobs."

100 Indeed.com, Broadband Consultant Job, accessed September 12, 2023, https://www.indeed.com/q-broadband-consultant-jobs.html?vjk=9643964fd68c4a49

101 ZipRecruiter, "10 of the Most Popular Types of Broadband TV Jobs."

102 Indeed, "Broadband Cable Technician/Installer," accessed September 12, 2023, https://www.indeed.com/q-broadband-digital-installer-jobs.html?vjk=02bfabd5e3453bc9.

car can go anywhere a traditional car goes and do everything that an experienced human driver does."[103]

Autonomous vehicles could make highways and roads safer and more efficient by reducing human error. Advanced sensors and communication technologies are designed to help vehicles better navigate roads, improve traffic flow, and avoid accidents. While some jobs may be replaced by self-technology, there will also likely be new opportunities for skilled workers in several arenas of the autonomous vehicle industry. All AV arenas will need bright, talented, well-educated individuals to stay current with advancements in these fields. The exact nature of these changes will depend on a variety of factors, including the pace of adoption, the types of vehicles that become popular, and the ways in which policymakers choose to respond to these developments.

Jobs in the Autonomous Vehicle Field

Job Title	National Average Salary	Job Description
Perception Software Engineer	$169,200/yr or $81.35/hr[104]	This job role offers the challenge of developing cutting-edge tech and machine learning models for self-driving cars. A self-driving vehicle must be aware of objects, stationary or moving around it, and precisely self-navigate around them. This tech relies on a diverse set of sensors like LIDARs, cameras, and radars. Companies like Waymo, Tata, and others who are in the autonomous vehicle development industry require researchers and software engineers to know computer vision and machine learning techniques for perception on self-driving cars.

103 Synopsys, "What Is an Autonomous Car?," accessed September 12, 2023, https://www.synopsys.com/glossary/what-is-autonomous-car.html.

104 Talent.com, "Perception Engineer Average Salary in the USA," accessed September 12, 2023, https://www.talent.com/salary?job=perception+engineer#:~:text=The%20average%20perception%20engineer%20salary%20in%20the%20USA%20is%20%24169%2C200,up%20to%20%24209%2C000%20per%20year.

Jobs in the Autonomous Vehicle Field

Job Title	National Average Salary	Job Description
Strategic Account Manager	$133,003– $175,273/yr[105]	The companies need automakers and skilled software engineers in abundance, but these companies also need someone to sell their products. That's why a strategic account manager is as important as an engineer in the company. An account manager supports the product that a company is about to introduce, in this case, an autonomous vehicle. This account manager will be helping with developing, defining, and executing the strategies that are adapted for gaining profit for the company. Companies like Molex—an electronics manufacturing company closely associated with Tesla—required substantial customer-facing time and relationship building within a customer organization, and a strategic account manager working closely with sales, marketing, business units, and cross-functional teams was helpful for the company.
Field Service Technician	$54,590– $67,990/yr[106]	A field service technician helps in testing, operating, maintaining, and evaluating a self-driving vehicle system. The field service technician reports to the head of the autonomous vehicle operator, who is responsible for ensuring and maintaining the safety of the vehicle during operations. The field service technician performs a series of tasks centered around customers like operations and maintenance training, installation supervision, on-site system commissioning, taking customer calls, and acceptance testing.
Industrial Engineer	$68,732– $82,034/yr[107]	An industrial engineer supervises the process and system for better material flows and development of tools and strategies. Industrial engineers mainly find out ways to make each process efficient where there is a production involved. Industrial engineers leverage their knowledge of industrial engineering, materials flowing, project management, order fulfillment processes, and WMS for making the process efficient.

105   Salary.com, "Account Relationship Manager Salary in the United States," accessed September 12, 2023, https://www.salary.com/research/salary/alternate/account-relationship-manager-salary.

106   Salary.com, "Field Service Technician, Entry Salary in the United States," accessed September 12, 2023, https://www.salary.com/research/salary/ alternate/field-service-technician-entry-salary.

107   Salary.com, "Industrial Engineer II Salary in the United States," accessed September 12, 2023, https://www.salary.com/research/salary/benchmark/industrial-engineer-ii-salary.

Jobs in the Autonomous Vehicle Field

Job Title	National Average Salary	Job Description
Customer Success Field Representative	$50,126/yr[108]	A customer success field representative is the one who supports customers and collaborates with the sales and marketing teams to facilitate additional sales on highly technical production. This personnel represents the company while maintaining a personal relationship with the customers.[109]

RENEWABLE ENERGY

As noted by Salary.com, "Renewable energy, often referred to as 'clean energy,' comes from natural sources or processes that are constantly replenished. For example, sunlight and wind keep shining and blowing, even if their availability depends on time and weather."[110]

As technology continues to advance, we can expect to see more innovative solutions that combine highway infrastructure and renewable energy to create a more sustainable transportation system. Solar and wind power, for instance, are areas being explored for their ability to power roadways and create sustainable electricity into the future.

---

108   ZipRecruiter, "How Much Does a Customer Success Representative Make?," accessed September 12, 2023, https://www.ziprecruiter.com/Salaries/%20Customer-Success-%20Representative-Salary#:~:text=As%20 of%20Jul%2017%2C%202023,be%20approximately%%2020%2429.38%20an%20hour.

109   Sameer Balaganur, "Top Jobs In Autonomous Vehicle Industry & How You Can Land Them," AIM, accessed September 12, 2023, https://analyticsindiamag.com/ top-jobs-in-autonomous-vehicle-industry-how-you-can-land-them/.

110   Natural Resources Defense Council, "Renewable Energy: The Clean Facts," June 1, 2022, accessed September 12, 2023, https://www.nrdc.org/stories/renewable-energy-clean-facts#sec-whatis.

Jobs in the Renewable Energy Field

Job Title	National Average Salary	Job Description
Greenhouse Worker	$28,056/yr	A greenhouse worker tends to plants in greenhouses and ensures they receive proper care and growth. They may water and trim them or plant new flowers and other crops. Greenhouse workers may have the added responsibility of selling the plants they have cared for.
Recycling Worker	$37,935/yr	A recycling worker sorts recyclable materials, including glass, plastics, paper, wood, and concrete. They may also use heavy machinery to process items or separate them for proper sorting. Recycling workers also pay special attention to any hazardous materials regulations and follow safety protocols for machinery and operations.
Environ- mental Technician	$49,706/yr	An environmental technician is responsible for collecting samples that scientists later use for research. They may specialize in water, air, or soil and complete tests to understand and prevent pollution. This position may require individuals to travel and work in an outdoor environment to survey and gather data from a variety of areas.
Environmen- tal Scientist	$61,917/yr	An environmental scientist is responsible for conducting research to understand environmental issues and develop plans to fix them. They have a firm knowledge of the natural sciences, using it to develop waste-reducing programs, advise policymakers, and address heavily polluted areas. These scientists may perform research or develop experiments to analyze environmental factors and develop solutions.
Wind Turbine Technician	$63,301/yr	A wind turbine technician installs, maintains, and repairs wind turbines. They regularly inspect the equipment to make sure it operates correctly and produces enough energy. Wind turbine technicians require an understanding of the equipment's hydraulic, mechanical, and electrical systems so they can correctly diagnose problems.

Jobs in the Renewable Energy Field

Job Title	National Average Salary	Job Description
Urban Planner	$67,556/yr	An urban planner is responsible for designing plans that include homes, communities, land, transportation, and public utilities. Urban planners who work in renewable energy may develop these programs based on sustainable practices. These planners may work with engineers, builders, government agencies, and other clients to make and edit designs and combine a variety of environmental elements.
Solar Installer	$70,310/yr	A solar installer installs solar panels on the roofs of buildings, including family homes and businesses, or on land. They may also install ancillary equipment like batteries and controllers. Solar installers follow safety standards and procedures to protect the safety and accuracy of themselves, their coworkers, and clients during the installation stage.
Air Quality Engineer	$71,525/yr	An air quality engineer creates and maintains environments that deliver high-quality air to inhabitants. They may focus on indoor or outdoor air quality, and sometimes both, which includes working with emissions, ventilation systems, and contaminants. In the renewable energy field, they may work to create an efficient and natural air purification system and may collaborate with other energy experts throughout the process.
Energy Manager	$77,233/yr	An energy manager is responsible for reducing energy usage for an organization. They may regularly monitor usage to understand how and where the energy is used and how that usage affects business operations. Energy managers may develop sustainability programs and implement systems to control energy consumption.
Energy Engineer	$86,670/yr	An energy engineer helps buildings implement more energy-efficient systems and structures. Their work involves lighting, air-conditioning, air quality, and other home or building systems. Energy engineers usually operate during the construction or remodeling parts of building creation so they can design and implement these programs.

Jobs in the Renewable Energy Field

Job Title	National Average Salary	Job Description
Environmental Specialist	$89,179/yr	An environmental specialist assesses how a population affects the environment. They develop plans to prevent or correct common environmental issues based on their analysis of the samples they gather and surveys they complete. These specialists require advanced scientific and environmental knowledge and use observation and analysis skills.
Sustainability Engineer	$97,181/yr	A sustainability engineer designs and creates systems that adhere to certain sustainability specifications. They use mapping technology and different tools to create models and collect data that help them devise designs. They may work with other renewable energy experts to determine the best solutions and create or install renewable energy sources.
Environmental Health and Safety Officer	$98,587/yr	An environmental health and safety officer is responsible for implementing plans that focus on health and safety, including those that concentrate on wastewater, toxins, and hazardous materials. They regularly inspect the workplace and individuals' homes to make sure their space remains safe and free of health risks that may come from food poisoning, pests, disease, and more. These officers may travel to various environments or work in commercial and residential areas to perform inspections and implement solutions.
Solar Consultant	$114,207/yr	A solar consultant is responsible for assessing energy consumption and informing their clients of their usage. They also develop strategies that help businesses and families reduce their energy. Solar consultants may also sell solar products to clients.

INFORMATION TECHNOLOGY (IT)

According to career coach Biron Clark, "IT jobs are positions in the fields of computer software, hardware, data storage/retrieval, and computer support. Information Technology is a fast-growing industry that offers many high-paying jobs and career growth. It's also one of the best fields to find remote work … IT is such a broad industry that you can break into it with virtually any background … including without a college degree."[111]

Technology has led to job creation for over a century, but today's technology is accelerating at a pace never before seen in human history. Unfortunately, this billion-dollar industry has been largely outsourced to other nations. We need American businesses that specialize in every sector of the IT field so we can keep gold-mine IT businesses on American soil. The sheer scale of modern technology is staggering, and those with the skills to master this field will have continued opportunities, now and in the future.

Jobs in the IT Field

Job Title	National Average Salary	Job Description
IT Technician	$51,569/yr	An IT technician collaborates with support specialists to analyze and diagnose computer issues. They also monitor processing functions, install relevant software, and perform tests on computer equipment and applications when necessary. They may also train a company's employees, clients, and other users on a new program or function as well.
Support Specialist	$58,536/yr	Support specialists are responsible for reviewing and solving computer network and hardware problems for a business. They can work in a variety of industries to provide general support to a company's employees or at a technology or software-as-a-service (SaaS) company to provide technical support on user experience issues that require technical assistance.

111   Biron Clark, "What Is an IT Job? (Full Guide with Salaries)," February 13, 2023, accessed September 12, 2023, https://careersidekick.com/it-job-guide.

Jobs in the IT Field

Job Title	National Average Salary	Job Description
Quality Assurance Tester	$65,518/yr	Quality assurance testers are technicians or engineers who check software products to see if they're up to industry standards and free of any issues. This role is common for gaming systems, mobile applications, and other technology that needs further testing and maintenance when recommended.
Web Developer	$67,854/yr	Web developers design the appearance, navigation, and content organization of a website. They use coding languages such as HTML, CSS, and JavaScript to manage graphics, applications, and content that address a client's needs.
IT Security Specialist	$71,818/yr	IT security specialists work in various industries to build and maintain digital protective measures on intellectual property and data that belong to an organization. They help companies create contingency plans in case information gets hacked from their networks and servers. These professionals also create strategies to troubleshoot problems as they arise.
Computer Programmer	$73,218/yr	A computer programmer is someone who writes new computer software using coding languages like HTML, JavaScript, and CSS. Video game software can be updated to improve online gameplay, which is an opportunity for programmers to troubleshoot problems experienced by gamers after the game is released to the general public.
Systems Analyst	$82,373/yr	A systems analyst reviews design components and uses their knowledge of information technology to solve business problems. They identify ways that infrastructure needs to change to streamline business and IT operations. They can also assist technicians in training staff to implement the changes they propose.
Network Engineer	$89,326/yr	Network engineers work on the day-to-day maintenance and development of a company's computer network, utilizing their skills to make the network available and efficient for all employees within an organization.

Jobs in the IT Field

Job Title	National Average Salary	Job Description
Software Engineer	$93,817/yr	Software engineers apply their knowledge of mathematics and computer science to create and improve new software. They may work on enterprise applications, operating systems, and network control systems, which are all examples of software that can be used to help businesses scale their IT infrastructure.
User Experience Designer	$94,954/yr	A user experience (UX) designer is involved with all facets of product development regarding its purchasing, branding, usability, and functionality. They collect and review user feedback to determine what a product needs to be efficient, functional, and successful. They apply this feedback to the design, organization, and usability. These professionals then monitor the process of testing and revising products until they meet their consumers' high-quality standards.
Database Administrator	$98,860/yr	Database administrators employ specialized software to organize and keep track of data. The software can be associated with software configuration, security, and performance when applicable. These professionals frequently diagnose and solve complex IT issues related to the data infrastructure to ensure an organization's data is safe, accessible, and easy to navigate.
Data Scientist	$102,312/yr	A data scientist analyzes and organizes data to determine trends that can influence business decisions. Their methods and IT tools use statistics and machine learning to help collect and process a company's data such as financial records, sales, prospects, and lead generation. Some duties vary for specific industries. For example, data scientists in the healthcare industry keep electronic health records (EHRs) intact for hospitals to have access to confidential medical information. They may also use data to help healthcare organizations make sound business decisions.
Computer Scientist	$108,521/yr	A computer scientist applies their technological skills and resources to solve IT problems for businesses. They write new software to complete tasks in a quick and efficient period as well as develop new functions that can be of use for employees or clients. Some computer scientists may also be application developers who help program software to serve users. IT companies heavily rely on computer scientists to create new programming languages and bolster the efficiency of hardware and software programs.

Jobs in the IT Field

Job Title	National Average Salary	Job Description
IT Director	$111,971/yr	An IT director oversees the strategy and execution of IT operations for an organization. They ensure that department tasks align with the company's goals and development. These professionals may also collaborate with other internal IT professionals as well as executive management to generate contingency plans, budgets, and development goals.[112]

GPS AND GIS

"GPS is a system of 30+ navigation satellites circling Earth. We know where they are because they constantly send out signals. A GPS receiver in your phone listens for these signals. Once the receiver calculates its distance from four or more GPS satellites, it can figure out where you are."[113]

GPS is crucial to modern travel, and today's GPS is integrated into onboard automotive computers, smartphones, and toll monitoring. We can reasonably expect that tomorrow's GPS will be even more expansive. Being able to tell exactly where something happens is relevant to not only tracking movement but planning and building. This technology will be absolutely critical throughout the infrastructure rebuild.

---

112   Jamie Birt, "21 Different Types of IT Careers to Explore," Indeed.com, April 14, 2023, https://www.indeed.com/career-advice/finding-a-job/types-of-it-jobs.

113   NASA Science, SpacePlace, "How Does GPS Work," accessed September 12, 2023, https://spaceplace.nasa.gov/gps/en.

Jobs in the GPS (GIS) Field

Job Title	National Average Salary	Job Description
GIS Technician	$36,915/yr	A GIS technician creates, tests, and maintains the code for various GIS projects. They understand how to create process and edit geographic data to design maps and other features using GIS technology. Some GIS technicians may specialize in more advanced skills and techniques like spatial analysis, meaning the process of evaluating different layers of geographical data such as precipitation and buildings. Other responsibilities include managing geographic databases, analyzing satellite or aerial imaging, surveying land, and updating maps as needed.
CAD Technician	$50,073/yr	A computer-aided design (CAD) technician develops digital renderings of building design plans or machinery schematics. They work together with other professionals involved in constructing or manufacturing new buildings or machines, such as architects, engineers, and drafting technicians. Most CAD designers make their digital renderings of plans or schematics already developed by these other professionals. CAD technicians often create both two- and three-dimensional plans of the buildings or machines under construction. Some CAD technicians also construct design plans using three-dimensional printing.
Geospatial Engineer	$52,951/yr	A geospatial engineer gathers, evaluates, and maps geospatial data, meaning information related to a designated area's land, bodies of water, natural resources, and man-made features. They use a range of technology and strategies to perform their job duties, including satellite, aerial, electromechanical, and high-precision optical instruments and systems. Many geospatial engineers find employment opportunities in the army, where they specialize in analyzing unfamiliar terrain and determining if it's safe for troops to visit that site. Other geospatial engineers find employment in the private sector, such as at engineering firms, real estate agencies, or mining companies.

Jobs in the GPS (GIS) Field

Job Title	National Average Salary	Job Description
Crime Analyst	$55,135/yr	A crime analyst gathers, assesses, and interprets crime-related data to make educated predictions about potential future crimes. Crime analysts often work closely with law enforcement professionals such as police officers or forensics experts to analyze their region's crime statistics and develop better methods or programs to reduce crime rates. A crime analyst with a background in GIS can help law enforcement agencies develop or install surveillance technology to evaluate crime patterns in a particular area or stop potential crimes before they occur.
CAD Designer	$55,638/yr	A CAD designer creates digitized plans or schematics for new creations such as for buildings, cars, skyscrapers, bridges, or industrial machinery. CAD designers may develop these plans or schematics on their own or collaborate closely with other professionals like architects or engineers. Other responsibilities include developing timelines and budgets for each project, visiting project sites to evaluate progress, and staying up to date on the latest CAD news or developments.
Archaeologist	$59,252/yr	Archaeologists study historical artifacts and documents to learn more about ancient civilizations, animals, or land features. An archaeologist identifies potential sites where historical artifacts might be and takes part in excavations or digs. They then spend time analyzing these artifacts and using them to broaden public understanding of various factors related to the past. A background in GIS can help archaeologists locate potential excavation sites with greater speed and accuracy. Most states run a historic preservation agency where archaeologists can submit GIS data to a public mapping system of important cultural sites in the area.
GIS Analyst	$60,372/yr	A GIS analyst evaluates and translates the raw data provided by GIS tools such as satellites and remote sensors into maps and databases. Most GIS analysts begin their careers as GIS technicians and then advance into an analyst role after two or three years. They often oversee and help GIS technicians or interns, such as by developing and monitoring their workflows or internal quality standards. GIS analysts also use more complex assessment techniques and technologies than technicians do, such as by managing or updating relational databases. Other responsibilities include converting physical maps into digital ones, searching for patterns through spatial mapping, and creating new mapping tools.

Jobs in the GPS (GIS) Field

Job Title	National Average Salary	Job Description
Environmental Scientist	$61,065/yr	Environmental scientists study and evaluate various factors related to the natural sciences like water quality or pollution. Their goal is typically to identify and minimize the effects of environmental damage or human health conditions related to the environment. Most environmental scientists have a specialization, such as health and safety or restoration. Many professionals with a background in GIS become climate scientists or climate change analysts who focus on causes, effects, and mitigation strategies for climate change. Climate scientists analyze both historical and current climate patterns to predict future climate changes and help businesses or government agencies develop strategies for minimizing these changes.
Forester	$63,360/yr	Foresters oversee and maintain the land and natural resources in certain areas, such as for timberlands, parks, range lands, conservation sites, or forests. Their responsibilities often depend on the type of land that they supervise. For example, while foresters responsible for timberlands may spend much of their time analyzing the quality of timber, foresters who oversee national parks may spend more time conserving habitats for native animals or plants. Other duties may include measuring the quality of various natural elements like water or soil, helping with forest fire prevention or suppression efforts, and adhering to all local and federal regulations.
CAD Manager	$63,947/yr	A CAD manager oversees all operations and personnel related to CAD project teams. They handle tasks such as scheduling workflows, managing project budgets, evaluating the quality of their teams' designs, and identifying project resources. Many of a CAD manager's other responsibilities depend on the size of the team they manage and their industry, such as whether they specialize in manufacturing or construction projects.

Jobs in the GPS (GIS) Field

Job Title	National Average Salary	Job Description
Surveyor	$64,428/yr	Surveyors measure different topographical features such as land or bodies of water to determine property boundaries for personal, governmental, and commercial purposes. They may evaluate the property lines for new construction sites or to update the boundaries for existing areas. Surveyors use a range of GIS technologies to help them determine property lines, such as handheld GPSs, robotic total stations, satellite images, and aerial data. They can use the information gathered by this technology to analyze property lines based on various geospatial features and generate digital maps. Other responsibilities include researching past land or surveying reports, reporting their findings to clients, and helping create land or water deeds and other legal documents.
Urban Planner	$64,744/yr	An urban planner creates development strategies, plans, and programs for a specific region. Their goal is to fulfill the needs of that community, whether that means adjusting for population growth, assisting with economic challenges, building community, or aiding with social issues such as poverty. An urban planner might create these programs or techniques for a town, county, or another metropolitan area. They often work closely with professionals like local politicians or community managers. A background in GIS can help urban planners evaluate a region's crime rates, weather patterns, highways, and other geospatial features when creating development plans.
Cartographer	$66,082/yr	Cartographers create and update maps. Most cartographers today create digital maps using a range of GIS technology and strategies, including aerial photographs, data on precipitation patterns, satellite images, and ground survey results. Some cartographers might also use light-imaging detection and ranging (LiDAR) technology, meaning lasers attached to moving vehicles that evaluate topographical features. Cartographers might create maps for a range of reasons, such as for public use, city planning purposes, or national security reasons.

Jobs in the GPS (GIS) Field

Job Title	National Average Salary	Job Description
GIS Manager	$68,133/yr	A GIS manager oversees all operations and staff related to GIS products or plans. They handle responsibilities such as developing long-term strategies or goals, monitoring project budgets, hiring or training GIS professionals, and giving presentations or progress reports to key stakeholders. Some GIS managers may share a few job duties with GIS analysts, such as evaluating different types of geospatial data or creating digitized maps.
Application Developer	$80,272/yr	An application developer creates software for devices such as mobile phones, tablets, and laptops. Their goal is to design programs that function well and are easy for users to use. Application developers with a background in GIS may specialize in programs that provide users with accessible maps or GIS data such as weather forecasts. An app developer familiar with GIS may also incorporate GIS features into other types of applications that can track or monitor a user's location data.
Geographer	$82,076/yr	A geographer researches and evaluates how various features on Earth's surface develop or change over time. Their goal may be to observe natural alterations in various topographical features or evaluate how human influences, such as social movements or construction sites, have affected these features. They collect and analyze data from sources such as censuses, maps, aerial photographs, and satellite images. Geographers might also conduct surveys or scientific experiments.
GIS Developer	$93,375/yr	GIS developers create the software that runs and supports various GIS systems. They develop, test, and alter software code for various hardware systems that use GIS technology such as mobile phones and laptops. A GIS developer's efforts help ensure that features such as remote sensing work accurately and are easy for users to operate.

Jobs in the GPS (GIS) Field

Job Title	National Average Salary	Job Description
Technical Security Threat Intelligence Officer	$95,539/yr	A technical security threat intelligence officer collects and analyzes data related to past or potential cyberthreats, such as viruses or hackers. Their goal is typically to help organizations eliminate existing cyberthreats from their network or prevent these threats from occurring. Also called cyberintelligence analysts, technical security threat intelligence officers can find employment opportunities in both the private and public sectors. A background in GIS can help technical security threat intelligence officers identify the locations of hackers or the origin points of computing viruses.[114]

DRONES

"The term drone usually refers to any unpiloted aircraft. Sometimes referred to as unmanned aerial vehicles (UAVs), these crafts can carry out an impressive range of tasks, ranging from military operations to package delivery. Drones can be as large as an aircraft or as small as the palm of your hand."[115]

Drones have a wide range of applications, and their use is expected to continue expanding in the future. For instance, drones can play a role in monitoring and inspecting highway infrastructure. They can be used to quickly and efficiently survey roads, bridges, and other structures for signs of damage or deterioration. With their aerial perspective, drones can capture high-resolution images and videos, allowing engineers and inspectors to identify potential issues in a timely manner. Additionally, drones equipped with advanced sensors and imaging technology can

---

114   Indeed Editorial Team, "18 Types of Jobs You Can Get with a Degree in GIS," Indeed.com, updated June 24, 2022, https://www.indeed.com/career-advice/finding-a-job/types-of-jobs-with-degree-in-gis.

115   Sam Daley and Matthew Urwin, "Drones," Builtin, March 23, 2023, accessed September 12, 2023, https://builtin.com/drones.

help detect cracks, corrosion, or other defects that may not be easily visible to the naked eye. This early detection can help prevent accidents and ensure the safety of motorists.

Jobs in the Drone Field

Job Title	National Average Salary	Job Description
Photographer	$39,432/yr	Photographers and filmmakers use drone technology to capture images and scenes from new perspectives. Independent photographers capture unique images of landscapes, people, and events. Photographers also might work for journalism companies like magazines or newspapers to capture current events. These professionals often adjust lighting and angles and edit photographs using digital software. Drones can help photographers in many areas like disaster coverage, traffic reporting, and investigative reporting.
Roofers	$46,612/yr	Roofers install and repair roofs for commercial, public, and residential buildings. They might replace shingles, fill holes, cut roofing materials to fit their clients' needs, and install insulation in attics. This process often involves climbing on a ladder, taking manual measurements, and gathering a team of laborers. With drone technology, roofers can quickly assess any issues on a roof, which can be especially helpful with tall buildings and those with minimal access points. Some drone technology can also deliver materials to these high locations, providing a safer and quicker way to transport materials.
Firefighter	$48,617/yr	Firefighters are emergency response professionals who prevent and respond to fires and other safety situations. They often inspect fire safety of structures and locations; operate equipment like trucks, hoses; and ladders; and respond to emergencies within their assigned areas. Firefighters can use drones for several purposes. They might use them for inspections, like identifying emergency exit locations or inspecting locations like ceilings and behind walls that might be difficult to reach. Similarly, with emergency response, a firefighter might deploy a drone into unsafe situations to identify any potential victims and determine the best way to resolve the emergency.

Jobs in the Drone Field

Job Title	National Average Salary	Job Description
Lineman	$53,923/yr	A lineman installs, maintains, and repairs electrical wires connecting buildings and to power units. They often climb poles to inspect electrical boxes, identify problems with wires between power sources. and ensure safety compliance for all outside electrical equipment. Using drone technology, linemen and power line inspectors can inspect extremely high wires and wires connected through smaller corridors that they might have difficulty accessing.
Drone Pilot	$57,842/yr	Drone pilots are dedicated professionals who operate drones and other unmanned aerial vehicles. They may inspect drones to ensure safe operations, document drone findings, plan routes, and maintain and repair drone equipment. Drone pilots can work in several industries, including aerospace, agriculture, and military defense. Some positions may require technology experience that involves sensors, imaging, photo and video editing, or artificial intelligence.
Delivery Driver	$63,208/yr	Delivery drivers load cargo and transport it from one location to another. They may work for specific companies or delivery companies, where they often serve a specific location. For smaller packages, like mail, envelopes, and light boxes, delivery companies might leverage drone technology to deliver packages. As a drone operator in the delivery field, you might need to log package details, input customer location, and confirm deliveries.
Surveyor	$64,363/yr	Surveyors evaluate land and buildings to assist in new construction projects, land mapping, and property line determinations. Working with engineers, architects, and agricultural or urban planners, they collect data like property measurements and topography to ensure the size, property borders, and land types accurately match their records. Using tools like drones and GIS (geographic information systems), they can photograph and measure locations that might be difficult to access through standard means.

Jobs in the Drone Field

Job Title	National Average Salary	Job Description
Transportation Planner	$65,741/yr	A transportation planner is a professional who develops plans and programs for various transportation systems, including waterways, highways, and rail systems. They often work with engineers and architects, performing research on the most efficient way they might build or update infrastructure. Using drone technology, they can gather aerial data on land and existing transportation systems to evaluate how they might design and build new structures or fix existing ones. As transportation planners might also inspect existing infrastructure, they can use drones to survey different components like rail tracks, concrete beams, landslides, and bridge components.
Miner	$79,237/yr	Miners explore and excavate underground areas to retrieve minerals or oils or set up underground infrastructure. They might drive underground vehicles, use explosives, and operate equipment like pickaxes and construction vehicles to build, maintain, and explore underground systems. Drone technology can help ensure these processes happen safely, as they can photograph underground locations to determine what tools they can use, what paths they might follow and for what risks they might prepare.[116]

CRYPTOCURRENCY

As explained by Coursera, "Cryptocurrency is digital money that doesn't require a bank or financial institution to verify transactions and can be used for purchases or as an investment. Transactions are then verified and recorded on a blockchain, an unchangeable ledger that tracks and records assets and trades."[117]

---

116   Indeed Editorial Team, "9 Jobs in the Drone Industry," Indeed.com, accessed September 12, 2023, https://www.indeed.com/career-advice/finding-a-job/jobs-in-the-drone-industry.

117   Coursera, "How Does Cryptocurrency Work? A Beginner Guide," Coursera, updated June 15, 2023, https://www.coursera.org/articles/how-does-cryptocurrency-work.

When it comes to highway infrastructure, blockchain and cryptocurrency can have several potential applications. One example is the use of blockchain-based systems for toll payments and traffic management. Smart contracts are self-executing contracts with the agreement terms written directly into code; they can be used to automate toll payments and ensure transparent and efficient systems.

Cryptocurrency is increasingly relevant, especially to global travelers. Technologies are emerging that enable the tokenization of traveler identities and funds.

Tokenization

"In the blockchain ecosystem, tokens are assets that allow information and value to be transferred, stored, and verified in an efficient and secure manner. These crypto tokens can take many forms, and can be programmed with unique characteristics that expand their use cases. Security tokens, utility tokens, and cryptocurrencies have massive implications for a wide array of sectors in terms of increasing liquidity, improving transaction efficiency, and enhancing transparency and provability to assets."[118]

Additionally, blockchain can enable the tokenization of assets, including infrastructure projects. By tokenizing highway infrastructure projects, it becomes possible for individuals to invest in and own fractional shares of those assets. This can potentially help in unlocking funding for infrastructure development.

Although implementing such systems would require careful planning, cryptocurrency's value in highway infrastructure lies in the potential for blockchain-based systems to improve efficiency, trans-

---

118   Cryptopedia, "What Is Tokenization in Blockchain?," Cryptopedia, accessed September 12, 2023, https://www.gemini.com/cryptopedia/what-is-tokenization-definition-crypto-token.

parency, and ownership models in the management of infrastructure projects. This movement to microtransactions and single-source currency has the potential to generate meaningful careers, both for the self-employed entrepreneur and those who prefer working in a company environment.

Jobs in the Cryptocurrency Field

Job Title	National Average Salary	Job Description
Technical Writer	$58,292/yr	Technical writers create content about complex topics in a way that's easily understood. Technical writers often use previous work experience in a particular field to help translate complicated topics for things like brochures, manuals, and journal articles. Primary duties include conducting research, deciding the best way to present the information, and working with technical professionals to ensure accuracy and comprehensibility. Technical writers may also proofread and edit other writers' work.
Business Development Representative	$62,852/yr	A business development representative (BDR) is a sales professional who develops new business and customer relationships for their organization. Common duties include finding potential clients and building long-term relationships with them, using marketing campaign leads as sales opportunities, identifying client requirements, and suggesting products or services. Organizations working with cryptocurrencies employ BDRs to expand their client base and generate business.
Marketing Manager	$63,142/yr	A marketing manager establishes their organization's marketing policies and oversees the marketing campaigns. They may collaborate with advertising and promotional managers to plan events, analyze data to determine strategies, create estimates for campaign costs and potential sales, and evaluate the success of marketing campaigns. When working with cryptocurrency, marketing managers may also work with technical and financial teams when developing marketing strategies for the currency they use.

Jobs in the Cryptocurrency Field

Job Title	National Average Salary	Job Description
Web Developer	$67,562/yr	A web developer writes code for a website for both the user-facing front end and the information-supplying back end. They can work on constructing a new website or maintaining and updating existing websites. A web developer's primary duties include writing code within set parameters, working with web designers on site appearance, collecting and analyzing user feedback to improve user experience, updating and maintaining software, and ensuring security for websites and related applications.
Financial Analyst	$67,654/yr	A financial analyst makes investment performance predictions by gathering and analyzing data about current industry and financial environments. Their primary duties include making recommendations about investments, compiling and analyzing financial reports, meeting with management and executives to evaluate the organization's prospects, evaluating financial data to identify trends, and compiling and presenting financial reports. Financial analysts working with cryptocurrency also monitor market trends and currency values and incorporate that data into their analyses.
Data Scientist	$74,943/yr	Data scientists take data and translate it into easily understandable information and results. A data scientist's typical responsibilities include evaluating the best-fit model for the data, identifying solutions using raw data and machine learning, collaborating with other departments, collecting data, and communicating analytical results to organization executives. Data scientists often use statistics, programming, computer science, and research skills when performing analyses.

Jobs in the Cryptocurrency Field

Job Title	National Average Salary	Job Description
Cryptocurrency Trader	$80,081/yr	A cryptocurrency trader is a finance professional who uses cryptocurrencies to advance investments. The responsibilities for this role often include researching market opportunities, using data analysis and trading metrics to identify market inefficiencies and opportunities, determining how to improve risk management techniques, and developing, maintaining, and improving trading models and systems. Traders often work with other industry professionals to find trading opportunities and market data.
Product Manager	$96,377/yr	A product manager oversees the product development process from start to finish. They're often responsible for sales or product teams. Common duties for a product manager in the cryptocurrency industry include creating, following, and updating product road maps; prioritizing and assigning tasks for the development team; brainstorming with executives and stakeholders; writing feature descriptions for design and technical teams; and leading product prototype development.
Software Engineer	$115,723/yr	A software engineer develops software for businesses. In the cryptocurrency industry, this includes using programming languages like C++ and Python and working on blockchains, substacks, and analytical programs. Other responsibilities include designing and testing systems to specifications, documenting systems for future maintenance, managing systems with regular updates, recommending upgrades, and collaborating with other engineers and developers.

Jobs in the Cryptocurrency Field

Job Title	National Average Salary	Job Description
Machine Learning Engineer	$135,460/yr	A machine learning (ML) engineer creates artificial intelligence systems. They use statistics, programming, data science, and software engineering to create these systems. Their primary responsibilities often include designing machine learning systems, researching and implementing ML tools, studying and transforming data science prototypes, developing ML applications, and training and retraining systems as necessary.[119]

ALTERNATIVE MODES OF TRANSPORTATION

There is growing interest in alternative modes of transportation such as high-speed rail, which could provide an alternative to highway travel for long distances. Additionally, there are efforts to encourage the use of mass transit and active transportation such as biking and walking to reduce reliance on cars and highways. Below, we've listed the top ten alternative transportation methods, according to *How Stuff Works*. A truly unknown number of jobs might spider out from these categories, so we've taken the liberty of suggesting some job categories that may expand as these alternative methods become more and more the norm. The length of this section, again, reflects the sheer number of job categories that relate to the infrastructure rebuild.

---

119   Indeed Editorial Team, "How to Start a Career in Cryptocurrency (Plus Jobs)," Indeed.com, updated January 26, 2023, https://www.indeed.com/career-advice/finding-a-job/crypto-currency-careers.

Top 10 Alternative Transportation Methods	
By Jane McGrath, *HOW STUFF WORKS*	
Hybrid Vehicles	Train
Electric Vehicles	Mass Transit Rail
Alternative Fuel Vehicles	Bus
Car Sharing	Biking
Carpooling	Walking[120]

TECHNOLOGY-DRIVEN ALTERNATIVE TRANSPORTATION METHODS

It can't be emphasized enough that the full list of jobs related to all ten fields is too much to put into one book. Because technology really is directing our future, we will focus on five technology-driven job categories: hybrid, electric, and alternative fuel vehicles and also trains and mass transit rail.

Hybrid and Electric Vehicles

If any category of the transportation industry is poised to explode, it's the hybrid and electric vehicle industries. Let's look at the distinctions between these two very similar industries before we move on to the job lists for this field.

---

120   Jane McGrath, "Top 10 Alternative Transportation Methods," How Stuff Works, accessed September 12, 2023, https://science.howstuffworks.com/ environmental/green-science/10-alternative-transportation-methods. htm.

"What is a hybrid? Quite simply, a hybrid combines at least one electric motor with a gasoline engine to move the car, and its system recaptures energy via regenerative braking. Sometimes the electric motor does all the work, sometimes it's the gas engine, and sometimes they work together. The result is less gasoline burned and, therefore, better fuel economy."[121]

"What defines an electric car? In today's automotive landscape, an electric car is defined as a passenger vehicle that uses an electric drive motor for propulsion. This broad definition, which technically encompasses a number of powertrain setups, includes hybrid vehicles."[122]

Jobs in the Hybrid and Electric Vehicle Fields

Job Title	National Average Salary	Job Description
Customer Service Representative	$48,666/yr	Customer service representatives for an electric vehicle company act as the liaison between the client and the business. In this role, representatives respond to service requests, customer concerns, and overall feedback on electric cars. Customer service representatives often work directly with the clients as part of a larger team to ensure that the brand appears responsive to its customer base. They may also work with members of the team to schedule maintenance appointments for clients.

121   James Riswick, "What Is a Hybrid Car and How Do They Work?," *Car and Driver*, accessed September 12, 2023, https://www.caranddriver.com/features/a26390899/what-is-hybrid-car/.

122   Brendan McAleer, "Pros and Cons of Electric Cars," *Car and Driver*, accessed September 12, 2023, https://www.caranddriver.com/features/a41001087/pros-and-cons-electric-cars/.

Jobs in the Hybrid and Electric Vehicle Fields

Job Title	National Average Salary	Job Description
Hybrid Vehicle Technician	$52,335/yr	Hybrid vehicle technicians, or EV technicians, have work similar to that of a traditional technician or mechanic. These specialists repair or improve the functionality of a vehicle and work directly with the client's vehicles in a shop setting. Additional duties might include routine maintenance such as oil changes and inspections and more complex repairs directly to the electrical components of the vehicle. These technicians work with certain types and brands of vehicles, which earns them the title of EV technician.
Electrician	$56,205/yr	Electricians fill many diverse roles in the electric vehicle industry, including updating, maintaining, and installing charging stations and assisting in the wiring functions of the electric cars if they work on-site at the plant. These professionals may often fix and update service stations with new hardware and assist with maintenance of the vehicle's internal wiring system. Other job responsibilities may include following safety protocols, collaborating with a team, and ordering parts.
Machinist	$59,189/yr	Machinists create and install components of electric vehicles. These specialists operate machines such as grinders and lathes that automate the manufacturing process as much as possible and assist engineers in the formation and development of new solutions for electric vehicles. Additional duties may include machine cleaning, maintenance, following safety procedures, and ordering specialized parts. A professional in this role might develop and use specialized equipment that meets the unique needs of electric vehicles.
Sales Representative	$64,440/yr	Sales representatives help customers choose electric vehicles that meet their needs. They may also keep records of current and past clients, work directly with cold leads to convert them into prospective buyers, and work with other members of the team to secure appointments for sales consultations. These sales professionals may stay informed on industry news and vehicle changes to better understand how they can assist their clients.

Jobs in the Hybrid and Electric Vehicle Fields

Job Title	National Average Salary	Job Description
Urban Planner	$64,744/yr	Urban planners assist with the development of electric vehicle infrastructure and determine where charging stations may exist. Urban planning is part of what makes electric vehicle ownership possible by ensuring that they can integrate into public roadways. Additional duties include completing budget calculations, mapping sites for charging stations, and developing a strategy to ensure that placement and zoning are efficient for the maximum number of citizens.
Chemical Engineer	$82,042/yr	Chemical engineers serve a variety of roles in the field of electric vehicles. Their contributions depend on the type of fuel storage the vehicle uses. In certain types of electric vehicles, there may be chemical components that contribute to the car's overall function and power maintenance. Chemical engineers create chemical mixtures and solutions that enable the proper functioning of the vehicle. Additional duties may include helping other members of the team with new vehicle designs and development and updating the current solutions with relevant industry research and knowledge to generate better designs.
Mechanical Engineer	$82,786/yr	Mechanical engineers develop innovative solutions for the mechanical components of the vehicle. This may include conceptualization of parts such as pistons, engines, fuel transport systems, wiring harnesses, or other mechanical components of the vehicle. Additional duties include creating designs that adhere to current safety guidelines and working with team members to create mock-ups, plans, and developmental notes for vehicles. They may also assist in the troubleshooting process if technical concerns arise, or make special tools for that specific brand and type of vehicle for the public market.
Software Engineer	$116,182/yr	Software engineers create complex computational code and software that helps the overall function of the electric vehicle. Cars use different software and logic blocks in their computer systems that allow users to enjoy higher levels of amenities that are common in electric vehicles. Examples may include software that dictates the use of power under different vehicular conditions or software that accurately displays the car's mechanical audits (such as fuel levels and miles driven) to the user.

Alternative Fuel Vehicles

According to the EPA, "New models of both electric vehicles and plug-in hybrid electric vehicles are entering the market in increasing numbers each year. Other alternative fuel vehicles include those that run on compressed natural gas (CNG) or E85 (a mixture of about 85% ethanol and 15% gasoline)."[123]

The field of alternative fuel vehicle technology offers a variety of career options, from professionals creating the technology to the technicians who repair vehicles. As an alternative fuel vehicle, or AFV, technician, you would be concerned with diagnosing problems and providing repair and maintenance service on a new class of engines and components.

If you're interested in creating or refining alternative fuel technology, then you may wish to work in the field as a scientist or engineer. For example, if you want to improve the efficiency of lithium-ion batteries used in electric or hybrid electric vehicles, you might become a research chemist, exploring the theoretical basis of chemical reactions. Alternatively, in chemical engineering, you would provide a more concrete perspective in designing and testing components that create and house the chemical processes in the automobile itself. With an interest in the unique circuitry of electric vehicles, you might consider becoming an electrical engineer.

---

123   US Environmental Protection Agency, "Learn about Green Vehicles: What is a Green Vehicle?," accessed September 12, 2023, https://www.epa.gov/greenvehicles/learn-about-green-vehicles.

Jobs Related to Alternative Fuels

Job Title	National Average Salary	Job Description
Alternative Fuels Data Center Project Lead	$130,350/yr[124]	Provide vision and implementation strategy for projects related to sustainable mobility information. Support existing projects and help build new projects with clients. Balance scope, budget, and deliverables while considering the needs of website users, clients, stakeholders, developers, and industry partners. Required Qualifications: Relevant bachelor's degree and 9+ years of experience. Previous experience managing web software development. Passionate about mobility systems and reducing their impact on the planet. Able to turn concepts into actionable tasks and deadlines with budgets and timelines. Adept at identifying and resolving potential problems and blockers.[125]
Biofuel Engineer	$116,510/yr or $56/hr[126]	Understanding the job of a biofuel engineer requires first understanding biofuel: As opposed to fossil fuels—like oil, natural gas, and coal, which are formed by the earth over millions of years using ancient plant and animal matter—biofuels are created from organic matter like algae, corn, or even used vegetable oil from fast-food restaurants. While fossil fuels are nonrenewable and polluting, biofuels are renewable, nontoxic, and biodegradable. As a biofuel engineer, you're typically employed by universities, research labs, government agencies, and private energy companies, and it's your job to design and devise tools, processes, and procedures with which to generate biofuel—for example, ethanol and biodiesel—for the purpose of powering automobiles, heating homes, and even generating electricity.[127]

124   Glassdoor, "Alternative Fuels Data Center Project Lead," accessed September 12, 2023, https://www.glassdoor.com/job-listing/alternative-fuels-data-center-project-lead-national-renewable-energy-laboratory-JV_IC1148181_KO0,42_KE43,79.htm?jl=1008784520457.

125   Glassdoor, "Alternative Fuels Data Center Project Lead."

126   ZipRecruiter, "Biofuel Engineer Salary," accessed September 12, 2023, https://www.ziprecruiter.com/Salaries/Biofuel-Engineer-Salary#:~:text=How%20much%20does%20a%20Biofuel,%2Fweek%20or%20%249%2C709%2Fmonth.

127   Chegg Career Match, "What Does a Biofuel Engineer Do?," accessed September 12, 2023, https://www.careermatch.com/job-prep/career-insights/profiles/biofuel-engineer.

Jobs Related to Alternative Fuels

Job Title	National Average Salary	Job Description
Renewable Energy Engineer	$74,698-$100,994[128]	Renewable energy engineers build solutions related to energy use. Like all engineers, they use a combination of research, mathematics, design, and testing to create green solutions that generate energy without harming the environment.[129]

TRAIN AND MASS TRANSIT RAIL

Again, we come to a form of travel and/or commuting that is exploding in popularity for the environmentally conscious. We also come to another category where job opportunities are too numerous to count. The range of train and mass transit rail positions goes from engineers all the way to maintenance workers. Talascend puts out an *abbreviated* list of common railroad job titles. Take a look at the number of jobs on this list:

128   Salary.com, "Renewable Energy Engineer Salary in the United States," accessed September 12, 2023, https://www.salary.com/research/salary/posting/renewable-energy-engineer-salary.

129   DeVry University, "What Does a Renewable Energy Engineer Do?," September 29, 2021, accessed September 12, 2023, https://www.devry.edu/blog/renewable-energy-engineer.html#:~:text=Renewable%20energy%20engineers%%2020build%20solutions,energy%20without%20harming%20the%20environment.

Common Railroad Jobs

ENGINEERING (BUILDING LOCOMOTIVES AND ROLLING STOCK)

- Chief Mechanical Engineer/ Rolling Stock Engineer / Locomotive Superintendent
- Locomotive Engineer
- Civil Engineer
- Signaling Engineer
- Electrical Engineer
- Mechanical Engineer
- Project Controls
- Project Management
- Construction
- Permanent Way
- Track Engineering
- Quality Engineer
- Health, Safety, and Environmental Engineer
- Document Control Engineer
- Structural Engineering
- Telecoms Engineer
- Software Engineer
- Systems Administrator

TRAIN

- Engineer (Driver)
- Boilerman
- Chief fireman
- Conductor (Human Transport)
- Secondman
- Freight Carman
- Brakeman
- Guard

STATION

- Station Agent
- Stationmaster
- Porter

TICKETS

- Ticket Controller (Transportation)
- Revenue Protection Inspector
- Ticket Inspector

OPERATIONS

- Train Dispatcher
- Dispatcher/ Logistics Coordinator
- Crew Dispatcher
- Clerk
- Yardmaster
- Yardmaster's Trainee
- Freight Conductor
- Signalman (rail)
- Railroad Brake, Signal, and Switch Operators

MANAGEMENT

- Road Foreman of Engines
- Diesel Mechanic
- Apprentice Diesel Mechanic
- Freight Car Repairer
- Apprentice Freight Car Repairer
- Mechanical Service Operator
- Diesel Electrician
- Apprentice Diesel Electrician

Common Railroad Jobs

MAINTENANCE OF WAY

• Bridge Inspector	• Work Equipment Mechanic
• Gandy Dancer	• Apprentice Work Equipment Mechanic
• Length Runner	• Assistant Signal Person
• Railway Lubricator	• Bridge and Building Electrician
• Section Gang	• Bridge and Building Carpenter
• Signal Maintainer	• Mechanical Service Operator
• Track Inspector	• Track Laborer / Welder
• Traquero	• Utility Clerk
• Platelayer	• Electronic Technician
• Navvy (Navigator)	• Installation Technician[130]

For information on specific jobs related to cranes and mass rail, try searching one of the above job titles in your zip code. If education or training is required, don't give up! Instead, go after what you want, and allow yourself enough time to get there.

---

130 Talascend, "Railroad Jobs," accessed September 12, 2023, https://www.talascend.com/industry-focus/railroad-jobs.

PART 6

CONCLUSION

B-R-I-D-G-E THE GAP

CONCLUSION

I'd like to sincerely thank you for taking this book journey with me. It is my sincere hope that you gained valuable insight from the reading experience that will change your life for the better.

When we talk about bridging the gap, it seems both obvious that a gap exists, and it's hard to discern where it's really coming from. All in all, we've addressed seven areas throughout the book that address gaps we're working to "BRIDGE."

BEST TALENT
ROADWAY SAFETY
INSTRUCTION
DESIGN
GRADUATE AWARENESS
EQUALITY
SYMMETRY

THE GAP IN BEST TALENT

This book idea actually began with a need for personnel. Our office was neck deep in viable projects, and an appraiser was needed to finish nearly all of them. This should not have been a problem, but all of a sudden, there was a colossal struggle to find one.

"Stressful" doesn't even begin to describe the tension of that time period. There I was, running my own business; there was no one else to fall back on; I was ready to get it done; and I was stopped at nearly every turn because there weren't enough qualified individuals to do the work needed to close the job. People were willing to learn, but our deadlines were coming due within weeks. I couldn't train everyone needed in their respective fields, and if I could have, there wasn't the time.

Worse, as I look around at the overall transportation industry, I can see this is happening all over. A job deficit has already hit, and it's getting steadily worse. We need numerous bright individuals in the five careers needing immediate fulfillment, or we will be on a backward track of delays. Bridging the gap between the necessity to fix our highway infrastructure and the lack of qualified people in the industry is crucial for the development of our society and overall economic growth.

THE GAP IN ROADWAY SAFETY

In preparing my notes to conclude the book, the term that seemed to keep jumping off the pages was *safety*. First, I was reminded of the Fern Hollow Bridge collapse, as this was somewhat close to my home. Quite possibly, nothing in a motorist's experience is more terrifying than a bridge failure, and ultimately, the life, economy, and environmental tolls on a community can be staggering.

Postcollapse scene at the Fern Hollow Bridge in Pittsburgh, Pennsylvania.
(Photo courtesy of Pittsburgh Public Safety Department.)[131]

"On January 28, 2022, at 6:39 a.m. EST, the Fern Hollow Bridge collapsed. Nearby residents reported hearing a loud boom and a whooshing noise around 6:35–6:40 a.m. Many credited the early morning time of collapse for the lack of fatalities, as the bridge was a route for many school buses, PAT buses, and commuters to work."[132]

Less than one year after the Fern Hollow Bridge collapse, in the midst of writing about this very thing, a tragic train derailment occurred in East Palestine, Ohio.

As reported by NPR on February 16, 2023, "On Feb. 3 [2023], just before 9 p.m. ET, a Norfolk Southern train derailed near East Palestine, Ohio, a town of about 4,800 people near the border with Pennsylvania.

"As authorities work to assess the damage and investigate the derailment, more information has emerged this week about the

131 Denise Bonura, "Why the Fern Hollow Bridge Collapse Could Have Been Avoided," Pittsburgh Magazine, May 19, 2023, https://www.pittsburghmagazine.com/why-the-fern-hollow-bridge-collapse-could-have-been-avoided/.

132 Wikipedia, s.v. "Fern Hollow Bridge," accessed September 12, 2023, https://en.wikipedia.org/wiki/Fern_Hollow_Bridge.

chemicals in the rail cars, *a variety of contaminants and carcinogens. Some of the chemicals—five rail cars' worth of vinyl chloride—was* intentionally burned off in a 'controlled explosion' last week, which prompted a temporary evacuation of the area.

"Some residents have reported headaches and rashes in the days since the derailment. And many have expressed frustration at what they say is a lack of answers from the railroad company and public officials."[133]

The scene in East Palestine, Ohio. (Photo by Commissioner Tim Weigle, provided courtesy of Columbiana County Commissioner's Office.)[134]

Not only does a derailment send mammoth steel cars careening off track to cause possible injuries, but the cargo that some trains carry can be uniquely dangerous. It is seldom reported, but the water supply in this small town was severely compromised. Take a look at the damage that was still being reported in June of 2023, *four months* after the accident:

133   Becky Sullivan, "What to Know about the Train Derailment in East Palestine, Ohio," February 16, 2023, National Public Radio, Environment, accessed September 12, 2023, https://www.npr.org/2023/02/16/1157333630/east-palestine-ohio-train-derailment.

134   "Failed Wheel Bearing Caused Norfolk Southern Train Derailment in East Palestine, Ohio," National Transportation Safety Board, June 25, 2024, https://www.ntsb.gov/news/press-releases/Pages/NR20240625.aspx.

"Even though government and company officials have claimed the air is safe to breathe and the water is safe to drink, residents have continuously reported negative health effects from skin rashes, headaches, and dizzy spells, to nausea, diarrhea, shortness of breath and mouth numbness; farm animals, pets and crops have been contaminated.

"They are running out of drinkable water. Their children are understandably traumatized. They are still waiting on results for tests on their water supplies if they even got their water tested in the first place. There is so much more that is needed to begin the process of repairing the damage that has been done to this community." [135]

There is no telling the long-term damage to the environment this wreck may have caused. If cargo such as this is to be safely transported throughout our nation, infrastructure repairs to our railways can no longer be ignored.

These two incidents magnify the critical state our infrastructure is in. We can talk about road conditions and sustainability all day long, but when massive tons of steel and cement crash or cave in, it devastates communities and the environment we are working so hard to sustain.

A more reasonable approach to sustainability, safety, and economics is regular maintenance and investment into even the tiniest corners of our infrastructure. This begins when competent individuals of all races, colors, and creeds work together to take advantage of the BIL's financing, apply those dollars wisely and equitably, and adhere to a system of upkeep.

---

135   Maximillian, Alvarez, "East Palestine Residents Have Been Left Behind—and They're Running Out of Water," Real News Network, June 1, 2023, accessed September 12, 2023, https://therealnews.com/east-palestine-residents-have-been-left-behind-and-theyre-running-out-of-water.

THE GAP IN INSTRUCTION

If anything became clear to me during the writing of the book, it was the need for instructional resources for careers in the real estate sector. With this need facing our industry every day, I and my team developed a series of training modules geared toward the title and right-of-way fields.

Highway Infrastructure Courses Online focuses on providing students with a comprehensive understanding of the legal aspects involved in property ownership. These courses cover the following topics, with new courses being added soon:

Property Law	Title Research/Examination	Easements
Encroachments	Land Acquisition Procedures	Termination
Clear & Marketable Title	The Role of Title Insurance	Condemnation Procedures
Types of Easements	Negotiation Techniques	Enforcement Processes
Legal Framework for Land Acquisition	How to Conduct Title Searches/Examinations	Compensation Principles
Conflict Resolution Methods	Dispute-Resolution Strategies	Communication Techniques
Practical Insights in Negotiations	Ethics between Property Owners & Stakeholders	Relevant Legislation, Regulations & Case Law

Overall, Highway Infrastructure Courses Online aims to equip students with the necessary skills and knowledge to navigate the complex legal landscape of property ownership and land acquisition for transportation projects. My goal in the near future is to provide scholarships to individuals from underserved communities who are interested in one of the real estate sector career paths. By investing in infrastructure projects and providing opportunities for individuals to acquire the necessary skills and training, we can address the dire gap in instructional resources.

Highway Infrastructure Courses Online	highwayinfrastructurecourses.com

THE GAP IN DESIGN

Engineers and architects, with innovation, creativity, and conscientiousness, will bridge the egregious gap in design that imprisoned whole black and minority communities. Where infrastructure is collapsing, engineers will have the latest technological advancements in science to ensure long-term repairs. Where cement and smog settled for decades, architects will design green spaces with clean air and avenues to new businesses, schools, and medical facilities. Where populations have been boxed in concrete, both will collaborate to design beautiful bridges to close the gap of historical inequality.

To bridge the design gap effectively, collaboration between engineers, architects, and the communities affected by past designs is crucial. To prioritize the needs of underserved communities when redesigning highways, E/A teams who truly understand the damage caused by our current system will prioritize these three initial steps to ensure the future designs are more equitable:

1. Conduct Thorough Research

Organizing public consultations and workshops allows E/A teams to directly engage with community members and gather their input and feedback. This helps ensure that the redesign addresses the specific needs and preferences of the underserved communities. It's essential to actively listen to community members' concerns, ideas, and suggestions throughout the design process.

2. Promote Environmental Sustainability

When redesigning highways, E/A teams can prioritize sustainability by integrating green spaces, urban forestry, and environmentally friendly features. This not only enhances visual appeal but also contributes to the overall well-being of the community. Additionally, sustainable design choices, such as stormwater-management strategies and noise-mitigation techniques, can help minimize negative environmental impacts.

3. Prioritize Accessibility and Safety

Underserved communities often face challenges related to accessibility and safety. A main focus of E/A teams must be in designing highways that provide safe connectivity to important community amenities such as schools, hospitals, parks, and public transportation. This may involve improving pedestrian infrastructure, ensuring ADA compliance, and implementing traffic-calming measures to enhance safety.

Effective collaboration between E/A teams, urban planners, and other relevant professionals will drive projects forward. By working together, these groups can leverage their diverse expertise to create comprehensive and inclusive designs that address the needs of underserved communities. This multidisciplinary approach ensures that all aspects of the redesign process are considered, including technical feasibility, social impact, and community engagement.

THE GAP IN GRADUATE AWARENESS

It is my sincere hope that this book becomes a bridge to the gap in awareness of the five careers we've highlighted. High schoolers and college students, especially, should be made aware of these stable, constructive, upcoming job opportunities. A historic restructuring is about to take place—is already underway, in fact—and we need engineers, architects, ROW teams, appraisers, and title specialists before we can build it.

THE GAP IN EQUALITY

New technology will impact underserved communities in a way that bridges the digital divide and provides equal access to education, mobility, and economic opportunities. By ensuring affordable and reliable internet access, expanding digital literacy programs, and creating initiatives for technology inclusion, we can empower under-served communities to thrive in the digital age. This can lead to improved educational opportunities, reduced unemployment, and greater economic mobility for all.

The best way to ensure that underserved communities actually receive these benefits is for women and minority groups to be part of the teams creating new infrastructure and making decisions regarding it. For the rebuild to really address gaps such as lack of clean water, electricity, transportation, and education, we need representatives and advocates from those communities active throughout the rebuild.

Women and minority groups are the bridge to the gap in equity all over the transportation sector. As we've discussed, the BIL provides financial advantages for women- and minority-owned businesses. This presents unique chances for you to be an integral part of the upcoming building boom. Consider pursuing the education and/or licensing of one of our five careers needing immediate fulfillment to establish yourself directly in the path of this historic boom.

THE GAP IN SYMMETRY

It is my honest desire to point out and rectify blatant racism in our built environment. A thin line exists, however, between countering racism and promoting a reverse form of it—by simply excluding a different race of people.

When we bridge the gap in symmetry, we recognize that when people from different backgrounds collaborate and share their perspectives, it leads to a rich tapestry of ideas and solutions. By embracing diversity, we create an environment where new concepts can flourish, existing paradigms can be challenged, and innovative approaches can be developed. This drives progress and enables societies to address complex challenges with a broader range of insights and expertise.

Bridging the gap and symmetry is all about working together and celebrating what each culture brings to the table for the betterment of our shared environment. It's about acceptance that includes others who may have different concerns and perspectives. Without this symmetry, it's very likely that the rebuild could produce the same results we are fighting against, which is a society skewed to serve one people group over another.

VISIONS OF THE FUTURE

Bridging the gap is really all about the future. The evolution of transportation has brought us from simple canoes to space travel, and there's no telling where we could go next and how we will get there.

My vision of the future really is to see all the up-and-coming technologies we've discussed working in our daily lives. Imagine the transformation the world could see if we rebuilt our infrastructure utilizing all the most modern technology.

A typical day might include finishing that final report while your autonomous vehicle brings you to work. You are not concerned about your child's safety on their trip to school because your own autonomous vehicle will return from leaving you at work, pick the children up, and drive them directly to school. You have paid for the car with cryptocurrency, along with your drone, which supervises the children when they return home and can alert you to any dangers.

In the meantime, your AV will charge itself at a local charging station one mile from your home. When recharging is complete, the AV will stop at the grocery store, where a carryout service loads your order into the climate-controlled trunk you've unlocked with your phone.

The AV returns home, picks your daughter up, and takes her safely to karate class. It's waiting for you when you exit your job. You have a few extra minutes; maybe you'll program it to take the long way home so you can enjoy a catnap.

When you arrive home, the drone has already helped the kids with homework in an interactive, fun way that kept them occupied, learning, and completing their assignments. You are enjoying some family time and don't feel like preparing dinner, so you send your AV out to pick up the family's favorite meal. Once dinner is out of the way and homework is done, there's time to enjoy a family game, movie, or favorite activity. Technology has made a hectic life easier, safer, and more convenient, leaving time for the most important things.

• • •

I realize not all technological changes will unfold in such a picture-perfect way. Several potential challenges and limitations could arise with the widespread implementation of advanced technology, such as drones, AI, and geo-tracking (GPS). As technology continues to advance, it will be important for policymakers and regulators to address privacy concerns and establish rules regarding data collection and storage by advanced technologies and strike a balance between the benefits and potential risks associated with each advancement.

In other words, it's up to us to take these positions, participate in our local government, and be the ones to make change for our generation and those to come. What we as citizens and business owners do *right now* is what will create our future. In the end, it won't only be my vision that counts but also yours. I've pointed out the gap, but each individual who accepts the challenge of this boom will be the bridge.

www.ingramcontent.com/pod-product-compliance
Lightning Source LLC
Chambersburg PA
CBHW051723260326
41914CB00031B/1704/J